Methane - Innovations for a Sustainable Future

Edited by Sevcan Aydin

Published in London, United Kingdom

Methane - Innovations for a Sustainable Future
http://dx.doi.org/10.5772/intechopen.1008576
Edited by Sevcan Aydin

Contributors
Akinpelu Moronkeji, Aleksey Nikitin, Gamze Gürerk, Hamid Hashemi, Igor Sedov, Kun Cheng, Nomcebo Mthombeni, Paramespri Naidoo, Phakamile Ndlovu, Saeideh Babaee, Sevcan Aydin, Temidayo Daniel Adeniyi, Valeriy Savchenko, Vladimir Arutyunov, Weidong Chen, Yunlin Shao

First published in London, United Kingdom, 2026 by IntechOpen
IntechOpen is the global imprint of INTECHOPEN LIMITED, registered in England and Wales, registration number: 11086078, 167-169 Great Portland Street, London, W1W 5PF, United Kingdom

For EU product safety concerns: IN TECH d.o.o., Prolaz Marije Krucifikse Kozulić 3, 51000 Rijeka, Croatia, info@intechopen.com or visit our website at intechopen.com.

British Library Cataloguing-in-Publication Data
A catalogue record for this book is available from the British Library

Methane - Innovations for a Sustainable Future
Edited by Sevcan Aydin
p. cm.
Print ISBN 978-1-83635-466-6
Online ISBN 978-1-83635-465-9
eBook (PDF) ISBN 978-1-83635-467-3

Meet the editor

Prof. Sevcan Aydin is a Professor in the Biology Department at Istanbul University, Turkey. She earned her B. Sc. in Biology from Ege University and her Ph.D. in Environmental Biotechnology from Istanbul Technical University. Her research spans biotechnology, particularly anaerobic digestion, renewable energy, microalgal applications, bioremediation, food biotechnology, and gut microbial ecology. She has published 49 high-impact scientific articles (SCOPUS citations: 1754, h-index 23). Her work focuses on optimizing bioenergy from waste, detecting methanogens in anaerobic systems, and developing cost-effective microbial monitoring tools, contributing to sustainable and health-related biotechnological innovations.

Contents

Preface

Methane is best known as a powerful greenhouse gas that accelerates global warming, but it is also a highly flexible molecule that is important in energy systems, the chemical industry, and microbial ecosystems. *Methane – Innovations for a Sustainable Future* examines both the risks and opportunities it presents, helping us better understand where this molecule is right now. The chapters bring together recent ideas and methods ranging from toxicological impacts of methane on human health to low-emission methanol production schemes, and from gas-hydrate-based separation technologies to the structure and function of methanogenic communities in soils. What sets this book apart is that it does not just talk about reducing methane emissions. Instead, it puts methane at the centre of several areas, such as energy networks, carbon cycling, and nature-based practices. Throughout the book, methane is discussed as both a harmful part of the atmosphere and a useful resource for gas-separation systems, cleaner fuel production, and soil biotechnology. The book presents a coherent perspective on the potential roles of methane in sustainable futures by integrating chemical processes with ecosystem-level microbial dynamics. In that sense, it is not really a collection of papers on one topic; it is more like a map of new areas for research and practice.

Chapter 1 examines the gas hydrate approach, which aims to alleviate the energy and equipment burdens associated with the industrial-scale recovery of noble gases, covering both its theoretical underpinnings and practical details. While the high energy requirements of traditional cryogenic separation make selective separation of gases such as xenon, krypton, and argon, particularly when present in low amounts, difficult, the ability of hydrate cages to encapsulate certain gas molecules preferentially makes this method an attractive option. This chapter comprehensively discusses the temperature-pressure ranges that determine hydrate formation, how crystallization initiates and progresses, the role of additives in accelerating the process, and why the separation stage is so crucial to the resulting purity, all in light of current research. The limited availability of low-energy, safely scalable noble gas separation methods today makes gas hydrates a priority for both selectivity and efficiency. Therefore, the chapter provides a solid engineering framework that explains why next-generation separation technologies with low energy consumption will become increasingly important, while also providing a scientific basis for the sustainable gas management approach throughout the book.

Chapter 2 provides a comprehensive overview of the matrix conversion approach, which aims to transform natural gas-to-methanol production into a lower-emission process. This method, which reuses the heat generated during the partial oxidation of methane within the system, offers a unique design that reduces the external energy requirement while converting the feedstock into synthesis gas. The section explains how this basic structure can be expanded into a hybrid model that combines dry matrix conversion, in which CO_2 is directly incorporated into the process, with a steam reforming step. This integrated scheme allows reevaluation of heat generated at different stages of the process and the flexible adjustment of the gas composition required for methanol synthesis. Simulation-based flow charts and conceptual design steps demonstrate that the method is both industrially feasible and enables the chemical utilization of CO_2 alongside natural gas. Thus, the study positions matrix conversion not only as an efficient pathway for synthesis gas production but also as an innovative technology that invites a rethinking of carbon management in the gas-chemical industry. This perspective provides a strong foundation for making methanol production cleaner, more economical, and compatible with future carbon-constrained production approaches.

Chapter 3 examines how methane, along with other airborne pollutants, can affect human tissues, drawing on both clinical observations and microscopic pathology. Rather than treating methane solely as a greenhouse gas, the authors show that it also alters the chemistry of the air we breathe and contributes to the overall oxidative stress burden. When methane interacts with compounds such as ozone, fine particulate matter, or NOx, a chain of reactions is triggered that may quietly injure organs such as the brain, lungs, heart, liver, and kidneys. Throughout the chapter, the authors refer to a range of recent studies to show, often with quite detailed examples, how minor tissue injuries, detected through routine histopathology or various oxidative-stress indicators, may gradually contribute to various health problems. These include conditions such as asthma or vascular issues, and in some cases, even kidney damage and certain cancers. They also note that people in many low- and middle-income regions are more frequently exposed to these pollutants, so the health burden linked to methane and related compounds is not shared equally and can easily slip below the radar in everyday life. In doing so, the chapter shifts the discussion of methane beyond its climate implications and adds a much-needed biomedical dimension to the book's broader sustainability perspective.

Chapter 4 takes a broad look at how methanol can be converted into hydrogen, treating it not as a single reaction but as a system where several processes depend on one another. The authors describe how placing the reformer, the PEM fuel cell, the organic Rankine cycle, and the latent heat storage units within a single integrated system allows the considerable heat produced during steam reforming to be reused rather than wasted. In practice, the ORC section channels what would normally be excess heat into electricity production and, when conditions change, that stored heat can be returned to the system through phase-change materials. As a result, the overall efficiency rises noticeably. The chapter also points out that features such as on-site hydrogen production, direct conversion in a fuel cell, and gradual heat recovery from the exhaust stream make this arrangement both environmentally and economically appealing. The research investigates the correlation among cost, exergy efficiency, and emission reduction through multi-objective optimization techniques, and formulates a high-efficiency, low-carbon design framework for residential applications. The chapter's distinctive contribution to the literature lies in its integration of the reforming process with heat management and in its proposal of a cohesive structure that systematically mitigates thermal losses. The study bolsters the book's overarching vision by providing a solid foundation for sustainable hydrogen production from both engineering and theoretical perspectives.

Chapter 5 examines how the global climate crisis is reshaping the functioning of soil ecosystems and demonstrates that methane-producing archaea should no longer be viewed solely as producers of methane. Instead, they are presented as essential ecosystem engineers that help regulate carbon flow and maintain anaerobic energy networks. It walks the reader through the fundamental biochemistry of major methanogenesis pathways, then discusses how the mcrA gene has become a reliable marker for tracking these organisms. The chapter also provides a clearer picture of how archaeal communities differ across soil types and how their structure evolves over time. It also examines how these archaeal communities respond when the environment changes. Changes brought on by warming, prolonged dry periods that suddenly produce rain, or soils that consistently accumulate excess carbon dioxide all affect the structure of these communities in different ways. The authors explain how these communities reorganize and how these changes can alter the rate and direction of methane formation. Methanogens may help restore parts of the soil microbiome weakened by climate stress, underscoring their biotechnological potential. Their contributions to biogas systems, wetland and peatland rehabilitation, renewable biomethane production, and other climate-sensitive applications are discussed. A key objective of the chapter is to integrate the molecular details of methanogenesis with real-world strategies for soil and ecosystem management. Consequently, a holistic perspective emerges that reassesses methanogenic archaea as strategic actors bridging the gap between restoring degraded soil microbiomes, sustainable agricultural practices, and climate-smart energy production.

The studies collected in this book each explore their own disciplinary depth, yet when read together, they form a coherent path toward sustainability in what might be called the age of methane. We hope that the ideas, data sets, and design approaches presented here will inspire the development of new reactor concepts, the revision of toxicology guidelines, the shaping of more ambitious climate policies, and the creation of practices that place soil biodiversity at the centre of environmental decision-making. As methane continues to exert growing pressure on the climate system, our sincere wish is that *Methane – Innovations for a Sustainable Future* will offer researchers and students not only practical solutions but also new opportunities for collaboration as they work toward a more resilient and sustainable future.

Dr. Sevcan Aydin
Professor,
Faculty of Science,
Division of Biotechnology,
Biology Department,
Istanbul University,
Istanbul, Turkey

Chapter 1

Gas Hydrate Technology for Sustainable Separation and Purification of Noble Gases

Phakamile Ndlovu, Saeideh Babaee, Hamid Hashemi, Paramespri Naidoo and Nomcebo Mthombeni

Abstract

There is an urgent need for sustainable engineering in the processing sector, focusing on waste management, with chemical separation playing a key role in handling chemicals based on their physical properties. Noble gases, found in small amounts in the atmosphere, are essential for daily life, making their separation and purification crucial for practical use. Cryogenic distillation is traditionally used to separate noble gases from gas mixtures, but its high operational costs and high energy demands have prompted the search for alternative methods. This chapter reviews gas hydrate technology as an alternative method for separating and purifying noble gases, emphasizing its lower energy requirements. Details of various studies that report the use of gas hydrates with noble gases are presented, with the limitations of gas hydrate technology and future research directions. Overall, gas hydrate technology, due to its lower energy requirement, has the potential to be utilized in the separation of noble gases.

Keywords: noble gases, gas hydrate, gas storage, gas separation, phase equilibrium

1. Introduction

Noble gases, helium (He), neon (Ne), argon (Ar), krypton (Kr), xenon (Xe), and radon (Rn) are a set of gases that are chemically inactive or stable, meaning they are not prone to reacting easily with other substances. Their chemical stability results from having a complete set of electrons in their outer shell. Although these gases are rare in the Earth's atmosphere, their applications are widespread and continue to grow with the advancement of new technologies. Their use has been gradually rising and is now widespread across multiple areas of daily life, including medicine, technology, industry, and science. In the medical field, noble gases like xenon and argon are especially utilized in anesthesia and imaging [1]. These gases find other uses in space and scientific research, especially xenon, which is used in space exploration and radiation detection [2]. Moreover, noble gases are also important in environmental technologies such as Carbon Capture and Storage (CCS), where xenon and krypton can be used as tracer gases [3, 4].

Efficient separation and recovery of the noble gases from industrial gas mixtures is very important for applications such as medical purposes where high-purity gases are required. Moreover, the ability to selectively separate and store these gases could be valuable for space missions, where efficient storage and recovery of noble gases is essential as fuel for life support systems and certain propulsion systems [5]. Several techniques have been employed in the separation and purification of noble gases, with the selection of methods depending on the targeted contaminants. These techniques include the use of zeolites, polymetric membranes, cryogenic distillation, metal-organic frameworks, absorption, adsorption, and gas hydrate technology. Generally, separating industrial gases from gas mixtures is challenging because of the high capital costs related to energy consumption and the stringent purity requirements for the products [6]. Moreover, most of the available separation technologies suffer from the lack of regeneration of the materials used in the process, making them uneconomic to apply in continuous large-scale operations. Processes such as cryogenic distillation have complex equipment requirements such as cryogenic heat exchangers, liquefaction units, and compressors, which require specialized materials capable of withstanding low temperatures, leading to high costs [7]. A promising technique for stable, compact storage of noble gases is provided by gas hydrates. Gas hydrate technology has been proposed and/or investigated for potential use in processes such as carbon capture and storage [8], treatment of vinasse [9], the concentration of juices [10], desalination of seawater [11], methane gas storage [12], and subsea flow assurance [13]. In addition, the increasing interest in gas hydrates is driven by their large storage capacity, with one volume of methane gas capable of storing 160 volumes of gas at standard temperature and pressure (STP) [14]. This storage capacity surpasses that of compressed or liquid petroleum gases under similar conditions.

Gas hydrates are clathrate compounds created when small gas molecules are trapped within a three-dimensional water lattice, forming a solid structure similar to ice. The formation of gas hydrates depends on specific pressure and temperature conditions [15]. Hydrate-forming gases have their own unique phase diagrams (hydrate stability zones), which illustrate the temperature and pressure ranges where their hydrate is stable. **Figure 1** shows the phase diagram for the xenon-water system with a stable hydrate region at high pressures (0.153–9.91 MPa) and low temperatures

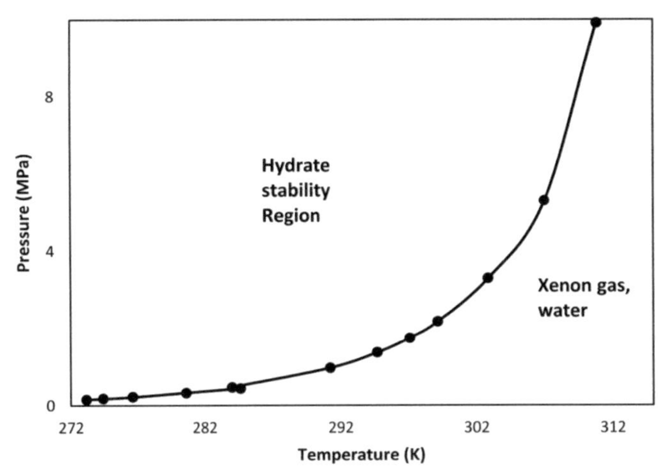

Figure 1.
Xenon temperature-pressure phase diagram [16].

(273.15–310.85) using the plotting range in the figure presented. This complicates the use of gas hydrate technology, as its stability conditions differ from those of room temperature and pressure. Additionally, an important factor in solid-liquid crystallization processes is nucleation, which is crucial for the formation and growth of gas-hydrate crystals. The time required for the first gas hydrate crystal to form is referred to as the induction time. This period is longer in homogeneous systems, making the gas hydrate formation process more time-consuming. The crystallization process takes place at the gas-water interface, meaning that interface-active agents like promoters and additives can have a greater impact on gas hydrate formation than bulk-active agents. The growth of gas hydrate at the interface is limited by diffusion resistance, which results in low mass transfer across the interface [17]. The gas hydrate crystalline has three unit cages that accommodate different gases; these are the sI, sII, and sH structures. The sI hydrate structure has two pentagonal dodecahedral (5^{12}) cages as well as six tetrakaidecaheral ($5_{12}6^2$) cages. The sII hydrate contains 16 pentagonal dodecahedral (5^{12}) as well as eight hexakaidecahedral ($5^{12}6^4$) cages, while the sH hydrate structure is formed by three pentagonal dodecahedral (5^{12}) cages, two dodecahedral ($4^35^66^3$) cages as well as one icosahedral ($5^{12}6^8$) cage [18]. Each unit cell of a gas hydrate can accommodate specific types of gases, but its storage capacity can be enhanced through the introduction of various additives.

Noble gases, such as xenon or krypton, can be safely and effectively stored at relatively low pressures and temperatures in hydrate frameworks [19]. In fields like medicine, industry, or science where noble gases must be transported or stored under strict control, gas hydrate technology shows potential solutions as it is solid and is stable within certain conditions of temperature and pressure. Gas hydrate technology has a low energy penalty, and the raw material requirements for the process are very low [20]. Furthermore, gas hydrate technology possesses a unique property of selective gas separation and removal over a range of pressures, temperatures, and pH conditions of the gas compositions. The encapsulation of gas molecules such as H_2, N_2, CH_4, CO_2, Xe, and other small gases in gas hydrates has proven to be an effective way of concentrating and separating various gases [14]. This technique can be applied to nuclear waste gas streams to separate the noble gases. This reversible technique is a potentially helpful technology for recycling or capturing gases from mixes since it enables the recovery of noble gases from the hydrate structure [21]. Moreover, gas hydrate systems serve to separate and capture these gases from emissions, helping to reduce the environmental impact of industrial processes. Current research focus can delve into ensuring process sustainability through the use of renewable energy sources as power sources for hydrate formation. Furthermore, techniques such as life-cycle assessments and techno-economic analysis can be carried out to give an overall and complete view of the process.

2. Separation and purification of noble gases

A radioactive form of Xe (^{127}Xe), released from the exhaust gases originating from nuclear fuel, is a key area identified as critical in achieving its separation from the gas mixture to attain sustainability in nuclear waste management [22]. Processes such as adsorption, membrane separation, absorption, and cryogenic distillation have been used to capture and separate radioactive gas streams containing iodine, krypton, and xenon. Krypton has been captured by cryogenic distillation, though this process needs pre-treatment steps involving the use of a scrubber to remove nitrogen [23].

The remaining gases containing Kr, Ar, H_2, Xe, O_2, and NO_x are then reduced to N_2 and H_2O by passing the gas through a rhodium catalyst at a temperature of approximately 900 K. The gas stream in contact with liquid nitrogen in the countercurrent absorption column results in the transfer of Kr, Xe, and Ar into the liquid nitrogen phase. This is a high-pressure process reaching up to 15 MPa, which poses safety risks due to the elevated concentration of radioactive materials. Exacerbated by the energy-intensive nature of distillation processes, which also incur high operational costs [24]. Another technique for krypton and xenon absorption is fluorocarbon absorption, where a precooled flue gas stream is bubbled through the fluorocarbon-12 (CCl_2F_2) with the Xe and Kr molecules being transferred to the liquid phase. This separation process has selectivity for Kr/Xe, resulting in higher gas waste volumes [25]. Furthermore, Nandanwar et al. [24] considered the treatment of the used nuclear fuel (UNF) gaseous radioactive nuclides, which include krypton, xenon, carbon, iodine, and tritium. The capture of the two volatile isotopes, iodine (^{129}I) and krypton (^{85}Kr), is of high environmental concern, as they are very detrimental to both humans and the environment.

Adsorption using solid sorbents has been in operation for a while. Metal-organic frameworks (MOFs) have been gaining attention due to their promising capacity. Experimental adsorbent data collection and vacuum pressure swing adsorption (VPSA) simulations were carried out for purifying and recovering xenon from semiconductor waste gas containing 0.1 mol% Xe and 99 mol% N_2. Adsorption of Xe was carried out by the use of a fluorine-doped metal-organic framework (MOF) containing copper (Cu) (FMOFCu) with the adsorption data collected through isotherm and breakthrough experiments using Xe and N_2. Experimental data was used in the gPROMS simulations, and approximately 99.99% of the Xe product was obtained through a two-stage VPSA process. Moreover, based on the economic and environmental evaluations, the proposed VPSA process was deemed environmentally and economically sustainable [26]. The use of different metal-organic frameworks (MOFs) are highly selective solids for the absorption of Xe, a significant semiconductor. Small-pore MOFs have been applied as well in the noble gas absorption and separation of Xe using microporous materials, and it was observed that optimal Xe-selective material must possess a pore size with the kinetic diameter of the Xe molecule. The absorption capacity of these MOFs was lower than the well-established MOFs as the separation potential was dependent on the channel shape and size as well as the interaction between the pore surface and the adsorbate [16]. In addition, activated carbon, a low-cost, stable solid with high surface area, has shown a capacity of approximately 50 wt.% at one bar and 298 K. Moreover, zeolites have been successfully utilized in the Xe/Kr selectivity. At relatively low pressures, an adsorption capacity of 520 ppm Xe and 60 ppm Kr was reported in a study using a zeolite-supported silver nanoparticle matrix under simulated nuclear processing conditions [24].

2.1 Fundamentals of gas hydrates as a method of separation

Gas hydrate technology has potential uses in the separation of gases since it utilizes different factors like molecular structure, size, temperature, and pressure. Hence, by manipulating these conditions, selective uptake in the gas hydrate can be achieved [27]. Khan et al. [28] utilized **Figure 2** to explain the mechanism through which gas hydrate technology is applied in the separation of gases. As depicted in the figure, a two-component mixture (HF-1 and HF-2; these are generic terms to illustrate the process) can be separated by exposing it to water and then adjusting the temperature

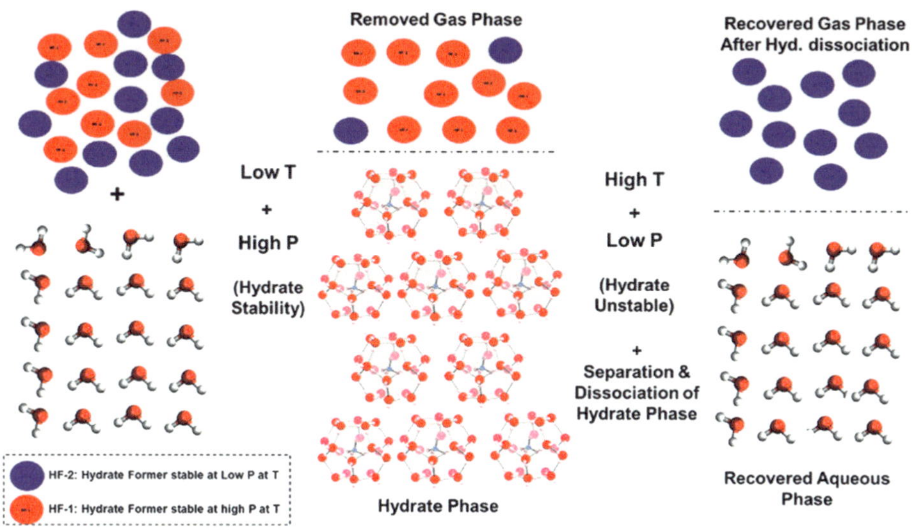

Figure 2.
Illustration of gas-hydrate based-gas separation process using two generic gases, HF-1 and HF-2 [28].

and pressure conditions to induce the formation and dissociation of hydrates. In this case, the pressure and temperature conditions were selected to ensure the uptake of HF-2, which preferentially forms the gas hydrate, thus getting trapped in the hydrate lattice. This leads to the separation of HF-2 from the gas mixture, resulting in the gas mixture being richer in HF-1. Thereafter, the solid gas hydrate was subjected to higher temperature and pressure conditions, leading to the dissociation of the solid gas hydrate, resulting in liquid water and gas phases comprising mainly of HF-2. Using this approach, it is possible for the two components to be taken into the gas hydrate phase; however, their concentration in the gas phase and the hydrate phase will be different, leading to their separation. In other applications, multiple hydrate formations and dissociations are necessary to achieve the desired separation of components. Therefore, using the hydrate technique, it is imperative for the temperature and pressure conditions to be selected carefully to achieve the desired or maximum attainable separation [28, 29].

2.2 Mechanisms of hydrate growth in gas separation

Hydrate crystallization occurs in two main stages (refer to **Figure 3**). The initial stage, called nucleation, involves the formation of a crystalline phase from molecules in a disordered or liquid state. The second stage, known as crystal growth, refers to the process where particles increase in size, resulting in the development of a crystalline structure [30]. The formation kinetics of hydrates in a gas mixture are influenced by both nucleation and growth processes. For effective use of gas hydrate technology in gas separation, one gas must be within its hydrate stability zone, while the others should be outside of it. In the presence of water, the gas within its stability zone does not instantly form a hydrate. Rather, the crystallization process relies on the mass transfer of components to the nucleation site and the chemical reactions occurring on the crystallization surface [31]. Hydrate nucleation takes place at a macroscopic scale, where tiny hydrate crystals form and expand until they reach the critical size

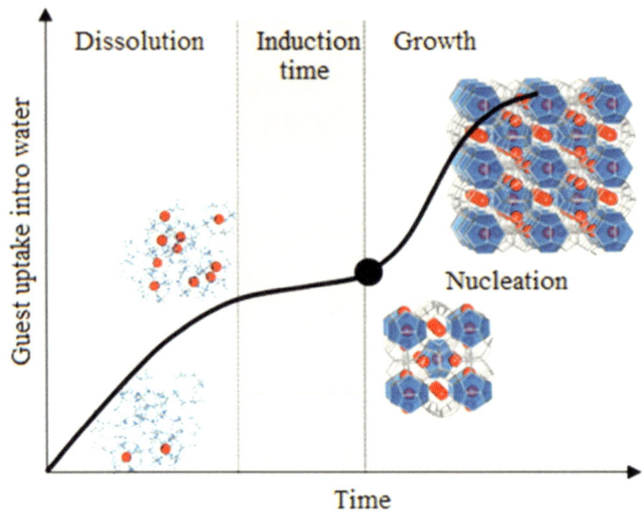

Figure 3.
Schematic illustration of hydrate growth sequence [30].

necessary for ongoing growth. During this process, subcooling occurs to overcome the surface energy required for hydrate formation. Impurities present in the system help facilitate both nucleation and subcooling. The formation of a larger crystal is referred to as the induction period, which is the time elapsed from when a system reaches conditions suitable for hydrate stability until the actual onset of hydrate formation [32]. Hydrate growth begins soon after the nucleation process. It is a combination of mass transfer from the bulk solution to the hydrate growth-liquid interface around the crystal and the reaction at the crystal surface. Overall, the growth is driven by the difference in fugacity of dissolved gas [33]. Utilizing molecular dynamics (MD) simulations, it was conclusively noted that gas dissolved in water and formed amorphous segments, which are precursors of the hydrate nucleation [34]. **Figure 3** illustrates the hydrate formation stages. Understanding these gas hydrate mechanisms is imperative in understanding the application of gas hydrate in noble gas separations.

Barrer et al. [35] performed a study on the energetics and kinetics of noble gases (xenon, krypton, and argon) occupying hydrate cage cavities, focusing on their formation kinetics, energy profiles, and equilibrium properties. The rates of formation differed among the gases, with xenon exhibiting the fastest hydrate formation, followed by krypton and argon. The energetic analysis indicated that xenon released the greatest amount of heat during formation due to its larger molecular size, whereas argon, being smaller, resulted in a lower heat release. The equilibrium findings demonstrated that xenon hydrates formed under the most extreme pressure and temperature conditions, while krypton and argon required progressively milder conditions. Additionally, statistical thermodynamic models of clathrate phases assume that the occupancy of guest molecules in the cavities is determined by the potential energy within fully formed cavities. However, it is suggested that clathrate hydrate forms via a surface reaction, where guest molecules occupy partially formed cavities. At the surface, some cavities only need one more water molecule to seal the entrance, preventing guest molecules from entering or escaping. These partial cavities have an equilibrium occupancy based on the potential energy for the guest. As water molecules complete the cavities, they integrate into the bulk phase, maintaining the same occupancy as before sealing [35]. Yang et al. [36] conducted a study on CO_2 hydrate formation in a multiphase flow regime for He/CO_2 and Ar/CO_2 under industrial conditions with high gas phase volume fractions and fluid velocities. The study found that the apparent rate constant for hydrate formation in the annular flow region was 0.45 s^{-1}, but this decreased significantly to 0.01 s^{-1} when the flow became slug or plug flow, highlighting the importance of good inter-phase mixing. The CO_2 hydrate formation was limited by heat transfer in the tubular reactor, especially under vigorous mixing conditions. Predictions based on kinetic and heat transfer analyses were consistent [37]. Heat transfer was found to be more important than kinetic rates when designing reactor dimensions. Using high thermal conductivity carrier gases like H2 can enhance heat transfer and improve hydrate formation. Additionally, adding trace impurities like C_3, H_2S, and C_2 can stabilize the CO_2 hydrate cage, reducing the partial pressure needed for hydrate formation at a given temperature.

2.3 Separation of noble gases using gas hydrate technique

The separation of two noble gases, xenon and krypton, is an industrial challenge, as these two gases have similar physical properties as well as closely sized molecules. An experimental investigation on the use of gas hydrate technology for the separation of xenon from krypton, argon, and xenon through a multi-stage process was conducted experimentally in a 52 cm^3 equilibrium cell. Using a gas feed mixture with 40.7 mol% argon, 25.7% xenon, and 33.6% krypton, it was observed that the concentration of xenon increased to 80.4% through the use of a two-staged hydrate formation and dissociation process. Moreover, it was noted that in comparison to the cryogenic distillation process, which requires very low cooling temperatures and variable pressures, the gas hydrate method for the present study reported a 20% energy cost reduction [38]. An experimental study to prove that gas hydrate technology was an excellent alternative to cryogenic distillation for the separation of the noble gas mixtures. The Kr and the Xe gas distribution in the resulting gas hydrate were identified using powder X-ray diffraction solid state, gas chromatography, and ^{129}Xe NMR spectroscopy. It was also observed that regardless of the Xe concentration in the feed vapor ranging from 3 to 100%, the sI hydrate selectively captured Xe from the feed mixture. This was possible even though Xe has a larger atomic radius of 1.08 angstroms compared to that of Kr, which is 0.88 angstroms. This might be due to the stability that a larger particle will bring to the hydrate structure [39]. These studies exhibit the capacity of gas hydrates to separate noble gases as well as increase their purity. In both an experimental and model-based study, the Newton numerical method was used to determine compositions of adjacent phases, a solid-based theory of van der Waals and Platteeuw with the Valderrama-Patel-Teja (VPT) equation of state, as well as the Non-Density-Dependent (NDD) mixing rules were used to predict the experimental data with a high degree of accuracy. Moreover, the experimental results from a single hydrate formation stage showed that the xenon was purified from 0.2 to 0.97 mole fraction from a mixture with argon [32]. In another computational study using molecular dynamics (MD) simulations, the final occupancy of the guest molecules and empty cages within 5^{12}, and $5^{12}6^4$, as well as the other cages, was investigated. Through the examination of the specific cages. The increase in the temperature and pressure improved the occupancy rate of Xe in both 5^{12} and $5^{12}6^4$ cages, while at elevated pressure occupancy of Kr in $5^{12}6^4$ cages was improved while the occupancy in 5^{12} cages was not affected by the changes in temperature and pressure. Overall, Xe preferentially occupied $5^{12}6^4$, while Kr occupied the 5^{12} cages. Moreover, the increase in the simulated pressure improved the occupancy from 0.37 to 0.41 at a simulation temperature of 260 K [40]. Additionally, this study demonstrated how the formation of Xe and Kr gas hydrates occurs (**Figure 4**). Temperature and pressure were varied to ascertain how they affected the inclusion rate and final occupancy. Beginning from the initial stage where liquid water does not contain any dissolved Xe and Kr molecules. The various stages of hydrate development are illustrated. These stages include the diffusion of gas into the liquid, the encapsulation of the gas molecules within the hydrate structure, and the eventual consumption of the liquid phase/gas phase, leading to the completion of hydrate formation.

2.4 Incorporation of additives in gas hydrate processes for noble gases

The additives, such as surfactants or promoters, can accelerate the kinetics of hydrate formation, leading to faster formation rates and larger hydrate volumes with lower energy requirements, ultimately allowing for more gas to be stored in the hydrate structure [41]. General hydrate promoters tend to reduce the hydrate formation pressure as well as increase the temperature, enhancing hydrate formation. Two types of promoters have been observed, one that does not alter the structure of the hydrogen-bonded networks, such as tetrahydrofuran (THF) and Sodium Dodecyl Sulfate (SDS), as well as the ones that are involved in the water networks, such as tetra-n-butylammonium (TBAB) [42]. Structural modifiers can further improve the stability of the hydrate, enabling it to hold a greater volume of gas. Additionally,

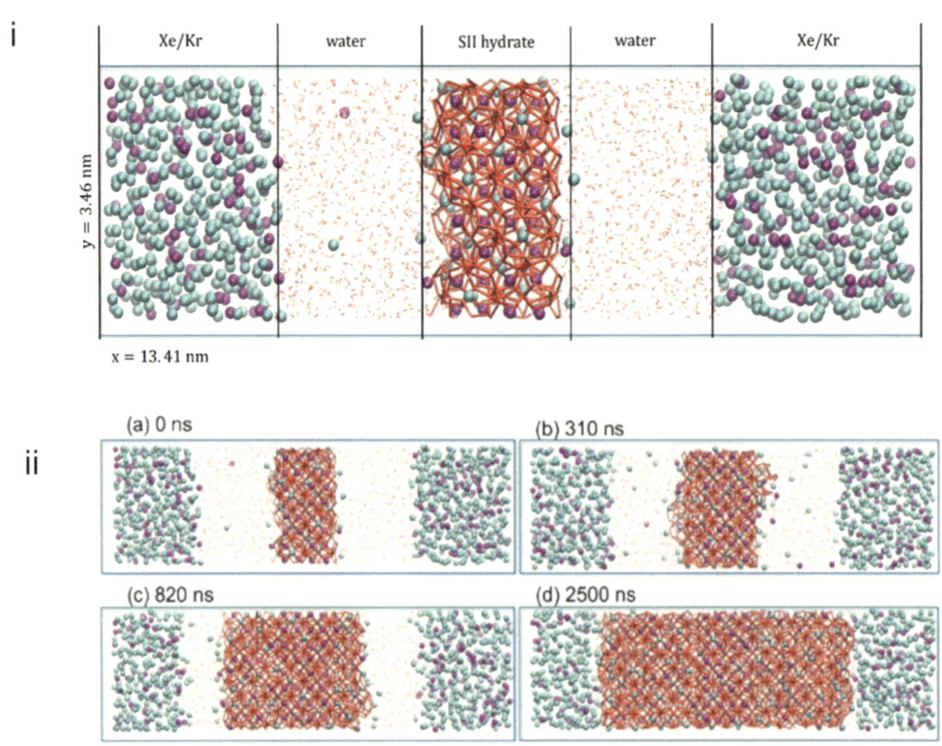

Figure 4.
(i) Configuration of the gas Xe/Kr gas mixture, sII hydrate, and water (ii) Snapshots of Xe/Kr hydrate growth at 500 bar and 260 K (a, b) gas molecules diffusing into the liquid leading to hydrate cage growth (c, d) and continuous growth of the hydrate [40].

larger guest molecules (krypton and xenon in the case of noble gases) can increase the gas storage capacity by allowing the hydrate to trap more gas within its crystalline structure, resulting in higher storage densities [43]. The use of dual-guest molecules can also further boost the overall storage capacity of the gas hydrate lattice [44]. The capacity to build gas hydrates with particular guest molecules (such as noble gases) can improve the separation process's selectivity. This process could entail fine-tuning the hydrate's lattice with additions or structural modifiers to selectively catch particular noble gases and maximize separation efficiency through a process known as hydrate lattice tuning [45]. The effects of quaternary ammonium salts (QAS) such as tetra-n-butyl ammonium acetate (TBAA), tetra-n-butyl ammonium fluoride (TBAF), tetrabutylphosphonium bromide (TBPB), tetrabutyl ammonium nitrate ($TBANO_3$) and TBAB on the stability conditions of CO_2 and Ar were assessed through experimental generation of hydrate dissociation data for the aforementioned systems. It was observed that the hydrate dissociation data for these systems with the Ar/CO_2 + water system showed that the quaternary salts could enlarge the hydrate stability zone for these gases [46]. In a separate study, Argon gas has been utilized in a CH_4-CO_2 replacement reaction as a secondary gas/co-encapsulation gas with the intent of improving the sequestration of CO_2 gas in the CH_4 gas hydrates. In this case, the dual-guest hydrate systems were used in order to improve the storage of or the amount of carbon captured in the hydrate structure, with both gases being encapsulated within the same lattice. The incorporation of Ar did not improve the total amount of CO_2 being stored in the hydrate lattice as the one obtained with N_2 gas [47].

In recent years, porous media materials such as silica sand and activated carbon, as well as clay, have been utilized in gas hydrate experimental work and have shown great potential in improving the storage capacity of the gas hydrate. The porous material utilizes the inter-tertial spaces where the gas and water interact on the pores and enhances gas hydrate nucleation and growth [48]. A theoretical study on the energy storage aspects was carried out on the molecular-level phase diagram of the clathrate hydrates on the methane and xenon gas hydrates. The results obtained agreed with experimental data, and self-preservation of gas hydrates was observed, which was linked to differences in thermal expansions [49]. The use of nanoparticles in hydrate research has shown promising outcomes in improving gas storage and accelerating hydrate formation kinetics. Due to their small size (less than 100 nm), nanoparticles significantly enhance mass and heat transfer between gases, the hydrate phase, and liquids. This improvement boosts the transfer of gas into the hydrate phase and increases the heat produced during hydrate formation, leading to a notable enhancement in both the gas volume consumed and the rate of hydrate formation [12, 50].

The noble gases in gas hydrates have been investigated in the presence of various additives and secondary gases to assess their effect on the hydrate phase equilibrium curve. **Table 1** gives a summary of the various studies and applications of noble gases in gas hydrate formation. It is evident from the tabulated studies that most investigations have focused on Xe and Kr gases, as these are the most commonly applied noble gases. Moreover, much focus has been placed on their phase equilibrium diagrams, as this area of study is very crucial in the uptake of the technology for various applications. It should be noted that gas hydrate technology has not been extensively utilized in industrial applications for the large-scale separation of noble gases. However, it has demonstrated potential in research and laboratory studies. The concept of employing gas hydrates for the separation of noble gases, including xenon, krypton, neon, and argon, remains a subject of ongoing investigation.

Reference	Noble gas/ Gas hydrate former	Study scale - Nature of study	Study outcome	Application of the study
[51]	Xe	Laboratory scale - Computational and experimental study on thermodynamic stabilization of Xe + large molecule guest substance (LMGS). Three LMGS used were neohexene (NH), methylcyclopentane (MCP), and methylcyclohexane (MCH) at a pressure of 0.1–1.0 MPa.	Observed that the equilibrium curve of Xe + MCH and Xe + NH hydrates shifted away from that of pure Xe hydrate under low pressure, while at high pressure it coincided with that of pure Xe. The equilibrium curve of Xe + MCP hydrate was superimposed on that of pure Xe hydrate throughout the pressure range.	Critical in the development of gas hydrate technology, noble gas capture, and storage.
[39]	Xe, Kr	Laboratory scale - Phase equilibrium of the Xe + Kr hydrate mixture for separation applications.	It was confirmed that the hydrate-based separation of the Xe-Kr separation process requires a temperature of around 250–273 K, which is higher than temperatures for cryogenic distillation (approx. **165.05 K**). Xe was trapped in the hydrate clathrate, producing Kr-rich vapors.	Potential use of hydrate-based Xe-Kr separation process as a replacement for cryogenic distillation.
[38]	Xe, Ar, Kr,	Laboratory scale - Experimental and simulation study on the separation of Xe from a gas feed mixture with Xe, Ar, and Kr.	The concentration of xenon was increased from 25.7 mol % to 80.4 mol % using a two-stage gas hydrate formation and dissociation process.	Separation of Xe from the noble gas mixture.
[47]	Ar/CH_4	Laboratory scale - Utilization of Ar to improve the CH_4-CO_2 replacement in the CH_4 gas hydrate.	When approximately 9.8 mol% Ar was incorporated into the replacement gas, it resulted in the CO_2 sequestration of approximately 16%.	Carbon capture and sequestration.
[52]	Ar	Laboratory scale - Study of the hydrate formation kinetics for argon in the presence of TBAB and SDS aqueous system.	Decrease in the induction time with the increase in the initial pressure while the consumption of the gas and the kinetic constant increase. The increase in the concentration of the TBAB led to an increase in the gas consumption and rate of hydrate formation with a decrease in the induction time. The addition of SDS at concentrations of 100 and 200 ppm at an initial temperature of 285 K and pressure of 6.1 and 8.1 MPa resulted in an increase in the induction time.	Kinetic improvement of gas hydrate formation.

Reference	Noble gas/ Gas hydrate former	Study scale - Nature of study	Study outcome	Application of the study
[53]	Kr	Laboratory scale - Experimental and modeling of hydrate equilibrium conditions for Kr + TBAB aqueous solution.	TBAB aqueous solution with concentrations of 0.2, 0.10, and 0.05 mass fraction had the most pronounced effect on the krypton hydrate, resulting in a shift of the P-T equilibrium line to higher temperature and lower pressure.	Shifting of Kr equilibrium curve for separation processes.
[54]	Kr	Laboratory scale - Phase transition of TBAB hydrates with Kr at 0.06–1.0 MPa and 280–290 K.	A new hydrate structure was observed in the higher-pressure conditions; with the Raman spectrum, it was confirmed that it was similar to pure TBAB.38H_2O.	Kr hydrate phase transition.
[55]	Xe	Laboratory scale - Experimental measurement and modeling of tetra-n-butyl ammonium bromide (TBAB) + Xe + water phase equilibrium data at a pressure of 10.09 MPa and temperature of 282.9–310.6 K on the hydrate equilibrium data.	At higher pressure, no observables were noted of TBAB on the xenon hydrate phase equilibrium. Good agreement was observed between experimental and modeling data.	Xe hydrate equilibrium diagram transition.
[56]	Ar	Laboratory scale - Experimental and thermodynamic modeling of hydrate dissociation conditions for the Ar + TBAB + water system. Experiments were carried out at 274–293.6 K and a pressure of 0.34–12.32 MPa.	The TBAB had a pronounced effect on the formation of the argon gas hydrate, which increased with the increase in its concentration. A thermodynamic model was developed to predict the hydrate dissociation of the argon + TBAB + water with good agreement obtained with experimental data.	Lowering pressure and temperature increase of the phase diagram for separation purposes.
[57]	Xe, Ar	Laboratory scale - Experimental phase equilibrium measurements for TBAB and TBAC hydrates with Xe and Ar as former gases at pressure values of up to 266–6114 kPa and temperature of 284–303 K.	At the investigated temperature and pressure ranges with the TBAB and TBAC with the two noble gases the required pressure was lower to form the pure gas hydrates.	Hydrate equilibrium phase transition.
[58]	Xe	Laboratory scale - Enclosing of Xe in tetra-n-butyl ammonium bromide (TBAB) hydrate -Xe-water system with TBAB mass fraction of 0.40 at 0.3 MPa and 283.6–287.4 K.	Using Raman spectral analysis, it was observed that no lattice vibration shift was observed with TBAB 26H_2O hydrate enclosing the XE while the lattice vibration of the TBAB 38H_2O enclosing the Xe shifted to a lower wavenumber.	Xe phase transitions to lower pressure and higher temperatures.

Reference	Noble gas/ Gas hydrate former	Study scale - Nature of study	Study outcome	Application of the study
[43]	Xe, Kr	Laboratory scale - Thermodynamic phase boundary for noble gas (Xe and Kr) gas hydrates at a pressure of 445 MPa and temperature (324.4–345.3).	With the Kr, a structural phase-transition point exists at 414 MPa and 390.20 K, while for Xe, hydrate structure sI did not exhibit phase changes over the region studied.	Structural phase transition.
[59]	Xe	Laboratory scale - The phase transition of the methylcyclohexane system from s-H to s-I structure at 281.5 K.	An overall enthalpy of s-H hydrate formation It was reported that neither cis-1,3-dimethylcyclohexane nor cis-1,4-dimethylcyclohexane has the capability of generating the s-H hydrate in the presence of Xe.	Structural phase transition.
[60]	Xe	Laboratory scale - Investigation of the thermodynamic stability of phase diagrams for the Xe hydrate system at a pressure of 70 MPa and a temperature range from 290 to 320 K.	It was observed that a three-phase curve coexists (hydrate + liquid Xe + gas), terminating at a critical endpoint of (289.76 K and 5.6 MPa).	Phase diagram thermodynamic analysis.
[61]	Xe	Laboratory scale - Investigated the composition dependence of thermodynamic properties of Xe hydrate.	Heat capacity in the temperature range 150–230 K, enthalpy of dissociation at 273.15 K, and pressure of 1 bar for hydrate clathrate of $Xe.29H_2O$.	Determination of physical properties.

Table 1.
Summary of experimental laboratory scale testing on noble gas equilibrium phase data and separation via gas hydrate studies.

3. Challenges and limitations

One main limitation and challenge in the application of gas hydrates in the separation process is the specific conditions under which the gas hydrate forms. This limits the range of conditions under which this technology can be applied. Another factor that affects the application of gas hydrates is the stability of the lattice structure formed with the noble gases. The stability of the gas hydrate is a critical parameter as it informs whether or not the structure formed will be able to accommodate the gas within the lattice structure. It must be noted that the stability of the gas hydrate is dependent on the temperature and pressure conditions; hence, the variation in these conditions leads to the dissociation of the gas hydrate. As shown in **Figure 1**, the stability zone for Xe gas hydrate exists at high pressure and low temperatures, making it challenging to achieve, as these conditions differ significantly from standard room temperature and pressure. The storage capacity of the gas hydrate is the main parameter essential for its utilization as a means of storing gases. Various additives and porous media materials have been utilized to attempt to increase the capacity of the gas hydrate to enable it to accommodate more gas. Currently, the Department of Energy has set a target of 180 cm^3 under 35 bars and 298 K using CH_4 gas. This parameter is defined as the volume of gas delivered per volume of the storage container and is also denoted by v/v or cm^3 (STP) cm^{-3} [62]. The other challenge with the application of gas hydrates in the separation process is the purity of the resultant gas. Contamination may result from incomplete separation during dissociation or the presence of other gases in the hydrate lattice, which would lower the process' efficiency and purity.

4. Future research

The gas hydrate research field is evolving and has great potential to be applied on a larger scale with improved efficiency as a cost-effective means of separating and storing noble gases. The key focus area that must be addressed is the integration of the gas hydrate method with the existing technologies. The hybrid can be created with technologies such as membrane filtration and cryogenic distillation to produce a process with various techniques for operation at varying temperatures and pressure conditions. In this instance, the gas hydrate technology can be utilized as a separation step while the cryogenic distillation or the membrane filtration can be a subsequent step for the purification of the gas. Moreover, this setup would be a scalable and energy-efficient solution for noble gas separation.

The use of emerging materials such as nanomaterials and nanostructured surfaces can be very essential in facilitating the nucleation process due to enhanced surface area, which improves the rate of gas hydrate formation. Moreover, these nanomaterials can serve to improve the selectivity and the stability of the lattice structure, hence improving the overall storage capacity of the gas hydrate clathrate. Porous materials, fluorocarbons, and metallic organic frameworks could be investigated as means of performance enhancement in the separation of noble gases. All these materials can be used as hydrate stabilizers or promoters for gas hydrate formation under mild or high temperatures and low-pressure regions.

It is also imperative to invest in research on the development of novel additives and guest molecules to enhance the optimization of the lattice structure for the accommodation of noble gases. In this case, larger molecules could be developed and incorporated into the hydrate structure to enhance the encapsulation of specific noble gases. These developments could enhance the selectivity of gas separation. Moreover, surfactants, structural modifiers, and co-solvents can serve to improve the selectivity and stability of the noble gas hydrates. All these additives will serve to stabilize the hydrate structure in temperature and pressure conditions outside the stability region for real-world conditions. This might entail developing low-cost materials for the hydrate formation and separation process, moreover optimizing energy usage as well as improving equipment efficiency.

5. Conclusions

The study of gas hydrates in the separation and purification of noble gases is still in its infancy. Notable strides have been made in experimental work on improving the kinetics and in the production of phase diagrams with the addition of various additives. These additives serve to shift the equilibrium curves to regions of higher temperature and lower pressure, hence serving as promoters to the hydrate equilibrium phase diagram. Simulation and modeling studies have paved the way for the assessment of various systems with conditions that would not be easily investigated experimentally. In industrial operations, gas hydrate can be applied in the separation of Xe and Kr from the nuclear waste stream, as it is a more economical process than the cryogenic distillation that is currently being applied. Future research should focus on scaling up existing studies, integrating them with emerging technologies, and developing methods to reduce induction time while enhancing the nucleation process. Overall, numerous possibilities are available in the application of gas hydrate technology for the separation and purification of noble gases. These research avenues need to be given research and industrial support, as outcomes from these investigations could be beneficial to the industrial players in improving the overall process of gas separation.

Conflict of interest

The authors declare no conflict of interest.

Author details

Phakamile Ndlovu[1]*, Saeideh Babaee[2], Hamid Hashemi[2], Paramespri Naidoo[3] and Nomcebo Mthombeni[1]

1 Department of Chemical Engineering, Durban University of Technology, Durban, South Africa

2 School of Chemical and Metallurgical Engineering, University of the Witwatersrand, Johannesburg, South Africa

3 Department of Chemical Engineering, Stellenbosch University, Western Cape, South Africa

*Address all correspondence to: phakamilen859@gmail.com

References

[1] Anna R, Rolf R, Mark C. Update of the organoprotective properties of xenon and argon: from bench to beside. Intensive Care Medicine Experimental. 2020;**8**(1):11. DOI: 10.1186/s40635-020-0294-6

[2] Di Gangi, Pietro. The xenon road to direct detection of dark matter at lngs: The xenon project. Universe. 2021;7(8):313. DOI: 10.3390/universe7080313

[3] Ju YJ et al. Noble gas as a proxy to understand the evolutionary path of migrated CO_2 in a shallow aquifer system. Applied Geochemistry. 2020;**118**:104609. DOI: 10.1016/j.apgeochem.2020.104609

[4] Weber UW. Applicability of Noble Gases for CO2 Capture and Storage Monitoring

[5] Pickett MT, Roberson LB, Calabria JL, Bullard TJ, Turner G, Yeh DH. Regenerative water purification for space applications: Needs, challenges, and technologies towards 'closing the loop. Life Sciences in Space Research. 2020;**24**:64-82. DOI: 10.1016/j.lssr.2019.10.002

[6] Duan Y, Li L, Shen Z, Cheng J, He K. Engineering metal-organic-framework (MOF)-based membranes for gas and liquid separation. Membranes. 2023;**13**(5):480. DOI: 10.3390/membranes13050480

[7] Bisotti F et al. Implementing robust thermodynamic model for reliable bubble/dew problem solution in cryogenic distillation of air separation units. International Journal of Thermofluids. 2021;**10**:100083. DOI: 10.1016/j.ijft.2021.100083

[8] Ndlovu P, Babaee S, Naidoo P, Moodley K. Kinetic studies of gas hydrates for CO2 capture in the presence of nanoparticles. Industrial and Engineering Chemistry Research. 2024;**63**(9):3867-3879. DOI: 10.1021/acs.iecr.3c04061

[9] Croeser N, Babaee S, Naidoo P, Ramjugernath D. Investigation into the use of gas hydrate technology for the treatment of vinasse. Fluid Phase Equilibria. 2019;**492**:67-77. DOI: 10.1016/j.fluid.2019.02.020

[10] Nkosi N, Nkazi D, Tumba K. A Review of Thermodynamic and Kinetic Studies Relevant to Gas Hydrate-Based Fruit Juice Concentration. Elsevier Ltd; 2023;**341**:111323. DOI: 10.1016/j.jfoodeng.2022.111323

[11] Nallakukkala S, Lal B. Seawater and produced water treatment via gas hydrate: Review. Journal of Environmental Chemical Engineering. 2021;**9**(2):105053. DOI: 10.1016/j.jece.2021.105053

[12] Ndlovu P, Babaee S, Naidoo P, Moodley K. Utilization of nanoparticles to improve the kinetics of CH4 gas hydrates in gas storage applications. Energy and Fuels. 2024;**38**(5):4480-4491. DOI: 10.1021/acs.energyfuels.3c04584

[13] Olajire AA. Flow assurance issues in deep-water gas well testing and mitigation strategies with respect to gas hydrates deposition in flowlines—A review. Journal of Molecular Liquids. 2020;**318**:114203. DOI: 10.1016/j.molliq.2020.114203

[14] Khokhar AA, Gudmundsson JS, Sloan ED. Gas Storage in Structure H Hydrates; 1998

[15] Koh CA, Westacott RE, Zhang W, Hirachand K, Creek JL, Soper AK. Mechanisms of Gas Hydrate Formation and Inhibition; 2002

[16] Simon CM, Mercado R, Schnell SK, Smit B, Haranczyk M. What are the best materials to separate a xenon/krypton mixture? Chemistry of Materials. 2015;**27**(12):4459-4475. DOI: 10.1021/acs.chemmater.5b01475

[17] Wang S, Yang M, Liu W, Zhao J, Song Y. Investigation on the induction time of methane hydrate formation in porous media under quiescent conditions. Journal of Petroleum Science and Engineering. 2016;**145**:565-572. DOI: 10.1016/j.petrol.2016.06.003

[18] Cai W, Huang X, Lu H. Instrumental methods for cage occupancy estimation of gas hydrate. Energies. 2022;**15**(2):485. DOI: 10.3390/en15020485

[19] Maclaughlin E, Umpleby R, Evarts J, Riley B, Thallapally P. Alternative Materials Literature Review A Review of Alternative Materials for Xenon Capture in Comparison to AgZ-PAN; 2023. Available from: http://www.inl.gov

[20] Kudryavtseva MS et al. Separation of CH_4 - CO_2 gas mixture by gas hydrate crystallisation: A parametric study. Gas Science and Engineering. 2023;**116**:205026. DOI: 10.1016/j.jgsce.2023.205026

[21] Loveday JS, Nelmes RJ. High-Pressure Gas Hydrates; 2008. DOI: 10.1039/b704740a

[22] Soelberg NR et al. Radioactive iodine and krypton control for nuclear fuel reprocessing facilities. Science and Technology of Nuclear Installations. 2013;**2013**(1):702496. DOI: 10.1155/2013/702496

[23] Elsaidi SK et al. Radiation-resistant metal-organic framework enables efficient separation of krypton fission gas from spent nuclear fuel. Nature Communications. 2020;**11**(1):3103. DOI: 10.1038/s41467-020-16647-1

[24] Nandanwar SU et al. Capture of harmful radioactive contaminants from off-gas stream using porous solid sorbents for clean environment–A review. Chemical Engineering Journal. 2016;**306**:369-381. DOI: 10.1016/j.cej.2016.07.073

[25] Wei Y et al. Efficient Xe selective separation from Xe/Kr/N2 mixtures over a microporous CALF-20 framework. RSC Advances. 2022;**12**(28):18224-18231. DOI: 10.1039/d2ra02768b

[26] Choi Y, Choi KM, Park K. Four-bed Vacuum Pressure Swing Dsorption for Xenon Recycling from Semiconduct Waste Gas; 2024. Available from: https://ssrn.com/abstract=4967782

[27] Misyura S et al. A review of gas capture and liquid separation technologies by CO_2 gas hydrate. Energies. 2023;**16**(8):3318. DOI: 10.3390/en16083318

[28] Khan M, Warrier P, Peters C, Koh C. Hydrate-based separation for industrial gas mixtures. Energies. 2022;**15**(3):966. DOI: 10.3390/en15030966

[29] Viswanadhan SK, Singh A, Veluswamy HP. Hydrate-based gas separation (HBGS) technology review: Status, challenges and way forward. Gas Science and Engineering. 2024;**131**:205465. DOI: 10.1016/j.jgsce.2024.205465

[30] Hassanpouryouzband A et al. Gas hydrates in sustainable chemistry. Chemical Society Reviews. 2020;**49**(15):5225-5309. DOI: 10.1039/c8cs00989a

[31] Ke W, Svartaas TM, Chen D. A review of gas hydrate nucleation theories and growth models. Journal of Natural Gas Science and Engineering. 2019;**61**:169-196. DOI: 10.1016/j.jngse.2018.10.021

[32] Sloan ED, Koh CA. Clathrate Hydrates of Natural Gases. Third ed. Vol. vol. 119. Boca Raton: Crc Press-Taylor and Francis Group; 2008. pp. 1-701

[33] You K, Flemings PB, Malinverno A, Collett TS, Darnell K. Mechanisms of methane hydrate formation in geological systems. Reviews of Geophysics. 2019;**57**(4):1146-1196. DOI: 10.1029/2018RG000638

[34] Zhang P, Wu Q, Mu C, Chen X. Nucleation mechanisms of CO_2 hydrate reflected by gas solubility. Scientific Reports. 2018;**8**(1):10441. DOI: 10.1038/s41598-018-28555-y

[35] Barrer RM, Edge AVJ. Gas Hydrates Containing Argon, Krypton and Xenon: Kinetics and Energetics of Formation and Equilibria. Available from: https://royalsocietypublishing.org/

[36] Yang D, Le LA, Martinez RJ, Currier RP, Spencer DF. Kinetics of CO2 hydrate formation in a continuous flow reactor. Chemical Engineering Journal. 2011;**172**(1):144-157. DOI: 10.1016/j.cej.2011.05.082

[37] Anderson G, et al. The Engineering Test Module (ETM): Construction, Commissioning, and Initial Hydrate Production Experiments. Los Alamos National Laboratory Report LA-UR-05-7377, Annual Progress Report Submitted to DOE/NETL. 2006

[38] Matizakurima F, Babaee S, Hashemi H, Naidoo P. Separation of Xenon from Noble gas mixtures of argon, krypton, and Xenon using gas hydrate technology. Industrial and Engineering Chemistry Research. 2023;**62**(36):14484-14496. DOI: 10.1021/acs.iecr.3c01174

[39] Chu H, Shin K. Highly-selective xenon–krypton separation using hydrate-based technology. Separation and Purification Technology. 2023;**319**:124094. DOI: 10.1016/j.seppur.2023.124094

[40] Liu L et al. A molecular dynamics study on Xe/Kr separation mechanisms using crystal growth method. ACS Omega. 2024;**9**(24):25822-25831. DOI: 10.1021/acsomega.4c00108

[41] Ganji H, Manteghian M, Sadaghianizadeh K, Omidkhah MR, Mofrad HR. Effect of different surfactants on methane hydrate formation rate, stability and storage capacity. Fuel. 2007;**86**(3):434-441. DOI: 10.1016/j.fuel.2006.07.032

[42] Lee YJ, Kawamura T, Yamamoto Y, Yoon JH. Phase equilibrium studies of tetrahydrofuran (THF) + CH4, THF + CO2, CH4 + CO2, and THF + CO2 + CH4 hydrates. Journal of Chemical & Engineering Data. 2012;**57**(12):3543-3548. DOI: 10.1021/je300850q

[43] Sugahara K, Sugahara T, Ohgaki K. Thermodynamic and roman spectroscopic studies of Xe and Kr hydrates. Journal of Chemical & Engineering Data. 2005;**50**(1):274-277. DOI: 10.1021/je0496692

[44] Song Y et al. Dependence of the hydrate-based CO_2 storage process on the hydrate reservoir environment in high-efficiency storage methods. Chemical Engineering Journal. 2021;**415**:128937. DOI: 10.1016/j.cej.2021.128937

[45] Kim DY, Park Y, Lee H. Tuning clathrate hydrates: Application to hydrogen storage. Catalysis Today.

2007;**120**(3-4 SPEC. ISS):257-261. DOI: 10.1016/j.cattod.2006.09.001

[46] Ilani-Kashkouli P, Hashemi H, Basdeo A, Naidoo P, Ramjugernath D. Hydrate dissociation data for the systems (CO2/CH4/Ar) + water with (TBAF/TBAA/TBPB/TBANO3 and Cyclopentane). Journal of Chemical & Engineering Data. 2019;**64**(6):2542-2549. DOI: 10.1021/acs.jced.8b01195

[47] Ndlovu P. Carbon Dioxide Encapsulation in Methane Hydrates. Durban: Ph.D, University of KwaZulu Natal; 2022

[48] Ndlovu P, Babaee S, Naidoo P. Review on CH_4-CO_2 replacement for CO_2 sequestration and CH_4/CO_2 hydrate formation in porous media. Fuel. 2022;**320**:123795. DOI: 10.1016/j.fuel.2022.123795

[49] Belosludov VR, Subbotin OS, Krupskii DS, Belosludov RV, Kawazoe Y, Kudoh JI. Physical and chemical properties of gas hydrates: Theoretical aspects of energy storage application. Materials transactions. 2007;**48**(4):704-710. DOI: 10.2320/matertrans.48.704

[50] Said S et al. A study on the influence of nanofluids on gas hydrate formation kinetics and their potential: Application to the CO2 capture process. Journal of Natural Gas Science and Engineering. 2016;**32**:95-108. DOI: 10.1016/j.jngse.2016.04.003

[51] Lee J, Yun S, Lee J, Lim J, Seo Y. Thermodynamic and computational approaches to the stability and structural transformation of Xe+ large-molecule guest substance (LMGS) hydrates. Journal of Molecular Liquids. 2024;**414**:126145. DOI: 10.1016/j.molliq.2024.126145

[52] Babaee S, Hashemi H, Mohammadi AH, Naidoo P, Ramjugernath D. Kinetic study of hydrate formation for argon + TBAB + SDS aqueous solution system. The Journal of Chemical Thermodynamics. 2018;**116**:121-129. DOI: 10.1016/j.jct.2017.08.030

[53] Babaee S, Hashemi H, Mohammadi AH, Naidoo P, Ramjugernath D. Experimental measurement and thermodynamic modelling of hydrate phase equilibrium conditions for krypton + n-butyl ammonium bromide aqueous solution. Journal of Supercritical Fluids. 2016;**107**:676-681. DOI: 10.1016/j.supflu.2015.07.030

[54] Jin Y, Kida M, Nagao J. Phase transition of tetra-n-butylammonium bromide hydrates enclosing krypton. Journal of Chemical & Engineering Data. 2016;**61**(1):679-685. DOI: 10.1021/acs.jced.5b00842

[55] Babaee S, Hashemi H, Mohammadi AH, Naidoo P, Ramjugernath D. Experimental measurements and thermodynamic modeling of hydrate dissociation conditions for the xenon + TBAB + water system. Journal of Chemical & Engineering Data. 2015;**60**(5):1324-1330. DOI: 10.1021/je5010092

[56] Babaee S, Hashemi H, Mohammadi AH, Naidoo P, Ramjugernath D. Experimental measurement and thermodynamic modeling of hydrate dissociation conditions for the argon + TBAB + water system. Journal of Chemical & Engineering Data. 2014;**59**(11):3900-3906. DOI: 10.1021/je500791e

[57] Garcia M, Clarke MA. Equilibrium conditions for TBAB and TBAC semiclathrates of xenon and argon.

Journal of Chemical & Engineering Data. 2014;**59**(11):3785-3790. DOI: 10.1021/je5006642

[58] Jin Y, Nagao J. Change in the stable crystal phase of tetra-n-butylammonium bromide (TBAB) hydrates enclosing xenon. Journal of Physical Chemistry C. 2013;**117**(14):6924-6928. DOI: 10.1021/jp310441u

[59] Shimada N, Sugahara K, Sugahara T, Ohgaki K. Phase Transition From Structure-H to Structure-I in the Methylcyclohexane + Xenon Hydrate System; 2003

[60] Ohgaki K, Sugahara T, Suzuki M, Jindai H. Phase Behavior of Xenon Hydrate System; 2000

[61] Handa YP. Composition Dependence of Thermodynamic Properties of Xenon Hydrate; 1986

[62] Mahmoud E. Evolution of the Design of CH_4 Adsorbents. Surfaces. 2020;**3**(3):433-466. DOI: 10.3390/surfaces3030032

Chapter 2

Low-Emission Production of Methanol on the Base of the Matrix Conversion of Natural Gas with the Addition of CO_2

Vladimir Arutyunov, Aleksey Nikitin, Valeriy Savchenko and Igor Sedov

Abstract

Ammonia and methanol production, which use natural gas to produce synthesis gas, play an important role in greenhouse gas emissions. This generates huge amounts of carbon dioxide, much of which is released into the atmosphere, thereby creating a greenhouse effect. One of the most effective ways to reduce CO_2 emissions is to use it as a feedstock for petrochemical large-scale production. In this work, we present a fundamentally new autothermal process for methanol production based on non-catalytic matrix conversion of methane, the main component of natural gas, and CO_2 into syngas. It shows the real possibility of utilizing a significant amount of CO_2 in the production of demanded petrochemical products and organizing gas chemical processes with almost zero CO_2 emissions. This is the most economically feasible approach to reducing the emission of this major greenhouse gas in technological processes.

Keywords: carbon dioxide, synthesis gas, methanol, natural gas, matrix conversion

1. Introduction

Methane is the most abundant gas source of energy and hydrocarbon feedstock on our planet, the vast resources of which can provide humanity for more than one century [1]. Since methane is in all respects an environmentally cleaner fuel than other hydrocarbons, its wider use facilitates the solution of global environmental problems [2]. Moreover, even in the distant future, it is the artificial production of methane and other hydrocarbons from carbon dioxide and water in abundance in the environment based on thermonuclear energy and known electrocatalysis processes that can ensure sustainable and carbon-locked production of hydrocarbon materials and fuels [3–5].

However, the use of methane as a feedstock for the production of a huge variety of modern petrochemicals is a much more difficult task than the use of traditional hydrocarbon feedstock. Since methane is the most thermodynamically stable hydrocarbon under normal conditions, its conversion into other less stable products requires high energy consumption and, as a rule, complex multi-stage technologies [6–8].

The situation is escalated by a growing problem of the global warming due to the increase of the atmospheric concentration of greenhouse gases such as carbon dioxide, which is considered to be one of the greatest environmental threats of our times. Anthropogenic carbon dioxide emissions come from combustion of carbon containing fuels and most technological and, first of all, chemical processes. Thus, the increased use of methane as a petrochemical feedstock should not be accompanied by a significant increase in energy consumption and CO_2 emissions.

Currently, the capture and utilization of CO_2 from the emission sources is considered as the main direction of solving this problem. Sequestration and subsequent use of carbon dioxide as a reagent could help mitigate its impact on the climate. Unavoidably produced CO_2 can be captured to be sequestered or be processed in many ways. CO_2 separation methods are absorption, adsorption, or membrane separation [9]. Adsorption of CO_2 is mostly used currently which is also called carbon amine scrubbing. Even though the underground CO_2 storage can be a large-scale capturing, sequestering in various forms other than underground storage is important in reducing CO_2 in the atmosphere because of its safety: leakage through tank or rock fissure can happen as a result of geological movement [10]. Therefore, it is preferable to involve sequestered CO_2 in technological processes.

Since it is very difficult to use extremely stable molecule of carbon dioxide as an oxidizer, for widespread use of methane as a petrochemical raw material, it is necessary from the very beginning to create a new generation of gas-chemical processes that ensure minimal energy consumption for methane processing and the ability to use in this processing the resulting carbon dioxide. These possibilities are opened up by non-catalytic autothermal processes based on partial gas-phase oxidation of natural gas [10]. The paper presents one of the most successful examples of such processes, demonstrating the possibility of producing such a high-demand petrochemical product as methanol with minimal CO_2 emissions. The technology uses the innovative principle of matrix conversion of natural gas [11], described in more detail in the next section.

2. Principles of matrix reforming

2.1 Principles of matrix reforming of methane

Matrix reforming is a type of non-catalytic gas-phase partial oxidation of hydrocarbons into syngas with internal heat recuperation from heated conversion products back into the stream of fresh reagents directly during the conversion [11]. This is one of the variants of the implementation of the so-called "superadiabatic" combustion [12], in which, due to the heat recuperation of the products, the temperature in the flame front can exceed the adiabatic burning temperature of a mixture of this composition, another example of which is filtration combustion [13].

The scheme of the matrix block is shown in **Figure 1**. In matrix reforming, recuperation of heat is realized by organizing flameless surface combustion, in which the combustion front is stabilized at a distance of 2–4 mm from the surface of a matrix made of heat-resistant material permeable to a mixture of hydrocarbon gas and oxidizer. Such materials can be perforated ceramics, foam metals, metal felt, or pressed metal wire. Recuperation of heat occurs due to intensive convective and (in the presence of a screen or in a geometrically closed matrix) radiative heat transfer between the flame

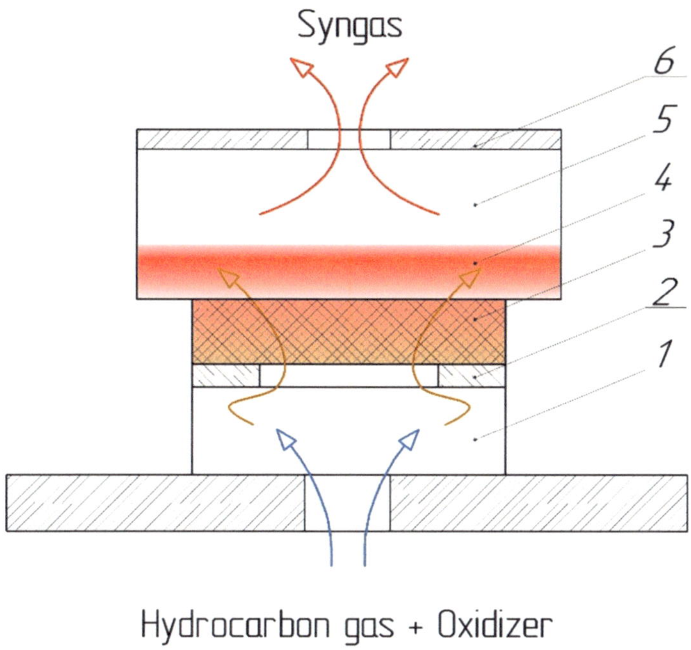

Figure 1.
The scheme of the matrix reformer block. 1 – preheating chamber; 2 – input aperture; 3 – permeable matrix; 4 – combustion front; 5 – reaction volume; 6 – radiation screen.

front and the solid matrix. As a result, the temperature of the outgoing conversion products is reduced by 600–700°C, and the working surface of the matrix is heated to 900–1000°C. Due to this, the fresh reaction mixture, passing through the matrix, heats up and enters the flame front already preheated. Preheating significantly expands the combustion limits of hydrocarbons and allows stable conversion of their very rich mixtures with an oxidizer, thereby significantly increasing the yield of syngas.

Heat recuperation from conversion products allows stable conversion of very rich mixtures of methane with atmospheric air, oxygen-enriched air, and oxygen at very low values of the oxygen excess coefficient α = [O2]/2[CH4], up to α = 0.30 [11]. However, the most complete conversion of methane and the maximum yield of syngas components H_2 and CO are achieved at values, which are optimal for this process. At lower values of α, the syngas yield decreases due to a decrease in methane conversion. The H_2/CO ratio, which is very important for the subsequent use of syngas, depend on the value of α and is in the range of 1.7–1.8 (α = 0.34–0.36). The process of conversion, with the exception of the thermal regime of the matrix, practically does not depend on the presence of nitrogen, the concentration of which in the oxidizer affects only the concentration of the components of the resulting syngas (**Figure 2**).

It is worth noting that with an increase in the concentration of oxygen in the oxidizer, the concentration of CO and hydrogen increases, not only due to a decrease of the nitrogen concentration in the synthesis gas but also due to an increase of the selectivity of their formation (**Figure 2**). When oxidized by atmospheric air, the following concentrations of syngas components are achieved: [H_2] ~23%, [CO] ~14%, [CO_2] ~ 3.6%. The conversion of both reagents is almost complete, over 95–97%. When oxidized with oxygen, similar values reach the values: [H_2] ~54%, [CO] ~28%, [CO_2] ~3.5%; oxygen conversion is complete, methane conversion is 84–88%.

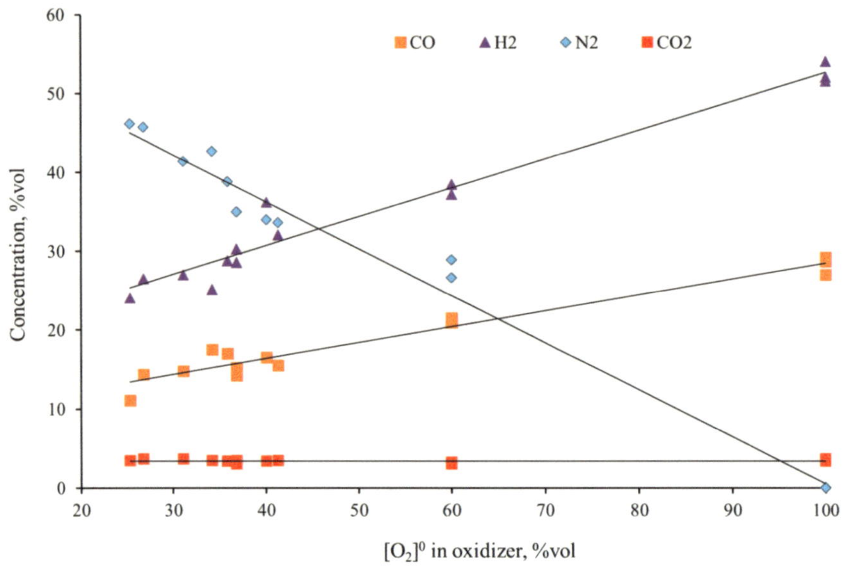

Figure 2.
Dependence of the concentration of main components of syngas on the concentration of oxygen in the oxidizer. α = 0.34–0.36.

2.2 Carbon dioxide (dry) matrix conversion of hydrocarbons

Kinetic and thermodynamic modeling of non-catalytic carbon dioxide conversion of methane under conditions close to matrix conversion demonstrated that CO_2 has practically no effect on the rate and mechanism of thermal pyrolysis of methane mainly into acetylene [14]. Under these conditions, CO_2, interacting with a hydrogen atom, was converted into CO and OH• radical. The accumulation of OH• radicals in the system contributed to the further conversion of the resulting acetylene. The stage of acetylene conversion in dry and steam reforming plays an important role. Because in this processes, acetylene is both an intermediate product of the thermal conversion of any hydrocarbon and a precursor of soot [15].

Thus, for any type of non-catalytic oxidative process, such as matrix conversion, oxidation in porous media or superadiabatic reactors, high-temperature pyrolysis of hydrocarbons, and others, where pyrolysis of hydrocarbons to acetylene plays a significant role, the addition of CO_2 to the reaction mixture can successfully solve a number of tasks, including CO_2 utilization, increasing the yield of syngas, and reducing soot formation [14].

Experimental tests of oxidative dry methane conversion in a matrix converter demonstrated, that, as expected, with an increase in the CO_2 concentration in the initial mixture, the matrix conversion temperature decreased until the process stopped. **Figure 3** shows the dependence of the temperature of the oxygen oxidative carbon dioxide conversion on the amount of CO_2 added to the initial mixture, relative to the concentration of methane, at various values of α.

At α = 0.4–0.5, even the addition of 50% CO_2 to methane practically does not lead to a decrease in the process temperature. With an increase in the methane content in the initial mixture (decrease in α), the conversion temperature decreased, and at α = 0.36, with the addition of CO_2 more than 20%, the conversion ceased.

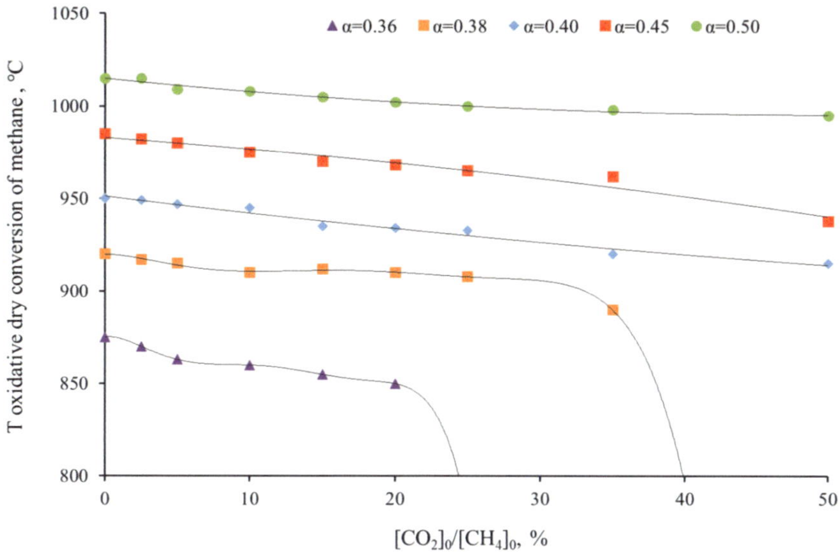

Figure 3.
Dependence of the temperature of the matrix conversion of methane and CO_2 on the concentration of CO_2 in the initial mixture at various α.

As noted earlier, the optimal conditions of matrix conversion of methane by oxygen are reached at $\alpha = 0.34$–0.36. This ratio of reagents ensures a stable mode of methane oxidation and maximum syngas yield. However, in order to involve a significant amount of CO_2 in the conversion process, it is necessary to increase the initial oxygen content in the mixture to $\alpha \geq 0.38$ (**Figure 3**). The **Figure 4** shows the concentration of the main syngas components of the carbon dioxide matrix conversion of methane at $\alpha = 0.38$.

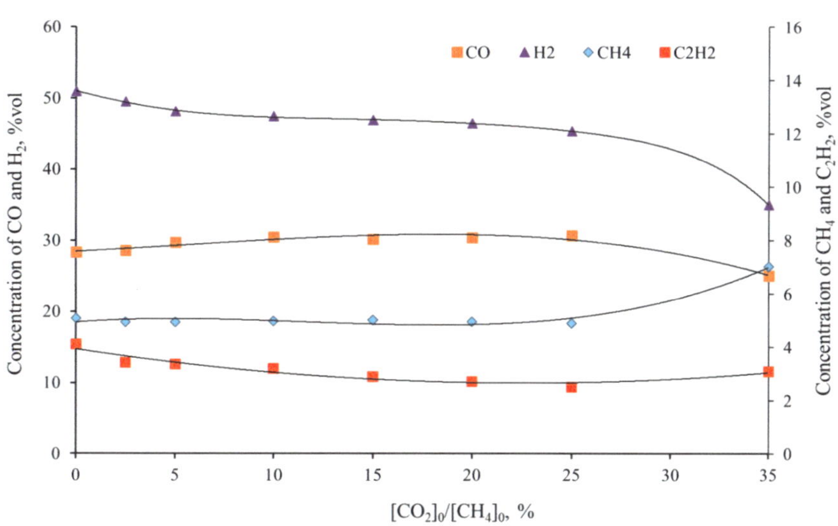

Figure 4.
Dependence of the concentration of syngas components on the concentration of CO_2 in the initial mixture at α = 0.38.

The increase of the CO_2/CH_4 ratio to 25% in the initial mixture leads to linear increase of CO concentration in the products, but at the same time the concentration of H_2 decreases. Despite the changes in the concentration of the main conversion products – H_2 and CO, the concentration of unconverted methane does not change. This may indicate that methane itself does not undergo dry conversion and the depth of its conversion is determined by the temperature of the process and the value of α. With a CO_2 content of more than 25%, a significant decrease in the process temperature was observed (**Figure 3**), which led to a decrease of methane conversion (**Figure 4**).

In contrast to the thermal carbon dioxide conversion of methane [14], in which a noticeable decrease of acetylene was observed with the addition of CO_2, in this case the acetylene concentration decreases slightly. The process temperature and the initial methane/oxygen ratio determine the acetylene yield, as well as the methane conversion. When adding CO_2 to the initial methane oxygen mixture H_2/CO ratio decreases from ~1.8 to ~1.5. An increase in the hydrogen content in the syngas can be achieved in the subsequent catalytic stage of steam conversion of the resulting CO.

Despite the demonstrated possibility of practical implementation of carbon dioxide matrix conversion of methane, the option of combined conversion using steam conversion, which is discussed in the next section, is more attractive.

2.3 Combined matrix and steam conversion of methane and CO_2

One of the problems of all reforming processes based on partial oxidation or dry methane conversion is the relatively low H_2/CO ratio, usually not exceeding 1.8 for partial oxidation and 1.0 for dry conversion [16]. Traditionally, this is solved by the subsequent partial catalytic steam reforming of part of the formed CO into hydrogen by the water gas shift reaction. The advantage of matrix reforming in this regard is that the process remains autothermal even taking into account the heat consumption for steam generation in a volume sufficient for the complete conversion the formed carbon monoxide into hydrogen.

The gas-phase nature of the matrix reforming makes it stable and easy to control, and the design of the converters is extremely simple and easily scalable. Compared with traditional catalytic steam reforming, matrix reforming has a very high specific volumetric capacity, since it proceeds at a distance of less than 1 cm compared to the length of tubes with catalyst in catalytic reformers, usually exceeding 10 m. This makes it possible to create compact low-tonnage installations with low gas capacity starting from several cubic meters per hour.

The high temperature of the syngas after matrix reforming makes it possible to add an additional amount of fresh natural gas and steam to it and direct the resulting gas mixture to the catalytic stage, in which, as the temperature gradually decreases from 950 to ~500°C, the endothermic processes of steam reforming of methane and partial steam reforming of CO to produce hydrogen sequentially proceed. The conceptual scheme of the reactor for the implementation of such a combined autothermal process is shown in **Figure 5a**. There is no need for a complete conversion of CO, since the remaining part of it is used in the subsequent synthesis of methanol.

Natural gas and CO_2 are preheated in a heat recovery boiler. Then, the first part of the natural gas is mixed with CO_2 and oxygen, and this gas mixture is fed to the reactor at the matrix conversion stage. The matrix conversion takes place at temperatures of 900–950°C and the second part of natural gas is added to the resulting synthesis

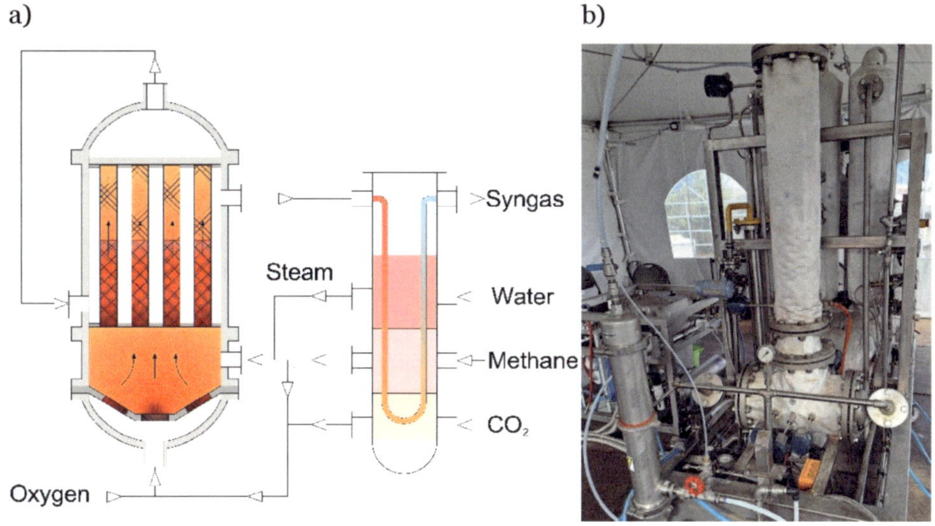

Figure 5.
a) Conceptual scheme of the autothermal reactor of combined oxygen matrix and catalytic steam conversions of natural gas and CO_2. b) Photo of the autothermal reactor of combined air matrix and catalytic steam conversions of natural gas.

gas along with water steam. Steam generation takes place in the first section of the heat recovery boiler. The natural gas/steam ratio is selected in such a way that not only the steam conversion of methane takes place but also the subsequent for steam conversion of CO formed at the matrix conversion stage.

The mixture of syngas, methane, and steam is then fed to the second part of the reactor – to catalytic steam conversion of methane and CO. The resulting syngas enters first into the intertubular space of the steam conversion section of the reactor, and then into the recovery boiler for heating the initial gases and generating steam.

The first series of tests conducted at the pilot plant (**Figure 5b**) showed that with combined air matrix and steam conversion of methane (without CO_2 additives), the temperature of the frontal catalyst layer was 900–920°C, and the upper catalyst layer was 500–550°C. The composition of the resulting dried syngas at a pressure of 12 bar, $\alpha = 0.38$, and a molar ratio of methane/steam = 3 is shown in **Table 1**.

The tests have shown that the combination of matrix and steam conversion of methane allows you to change the ratio of H_2/CO in the resulting syngas in a wide range. The main parameter affecting the hydrogen yield is the amount of steam supplied to the steam conversion stage. The syngas (without N_2) produced at the stage of combined oxygen matrix and steam conversion of methane, in addition to hydrogen, contains 25–30 vol. % of carbon oxides that can be used for the production of methanol and, thus, practically avoid their emission.

Component	CH_4	O_2	H_2	CO	CO_2	N_2
Stream III, vol. %	0.4	0	35.5	3.6	12.1	48.4
P = 10 бар, α = 0.38.						

Table 1.
Composition of dry syngas produced at the combined air matrix and catalytic steam conversions of natural gas.

3. Conceptual technological scheme of methanol production based on the matrix conversion of methane and CO_2

To obtain methanol, one of the most common multi-tonnage products of basic organic synthesis, it is required that the ratio of H_2/CO in the resulting syngas was at least 2. Earlier in [17], a conceptual technological scheme for the production of methanol from natural gas was proposed, including the stages of matrix conversion of methane and subsequent catalytic steam conversion to increase the hydrogen content in the syngas. The proposed combination of non-catalytic exothermic matrix conversion with endothermic catalytic conversion of methane and CO can also be used in the production of methanol from methane and CO_2 (**Figure 6**).

The material and energy balances of the proposed methanol production process from methane and CO_2 were calculated with the Aspen HYSYS program. To model the thermodynamic properties of the system components, we used the Peng–Robinson method, which is widely used for such systems [18]. The process of matrix conversion of methane is determined by detailed kinetics of gas-phase reactions [11, 14, 15]. Aspen HYSYS does not adequately describe the kinetics of non-catalytic partial oxidation process. For this reason, the stage of matrix conversion was not calculated, and the composition of the syngas of matrix conversion of methane and CO_2 was set manually, based on the results of earlier experimental studies. As it was shown earlier, matrix conversion at $\alpha < 0.38$ does not allow to involve a significant amount of CO_2 in the process, therefore, the calculation of the technological scheme was carried out

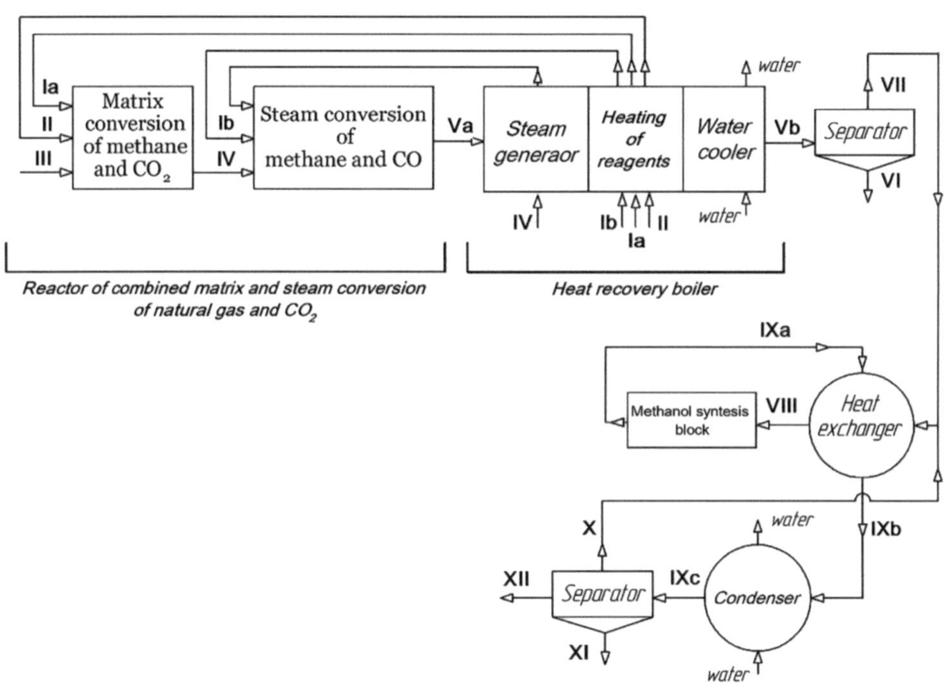

Figure 6.
The scheme of the methanol production process from methane and CO_2. Streams: Ia, Ib – natural gas (methane); II – CO_2; III – oxygen; IV – raw water; IV – synthesis gas after matrix conversion; Va, Vb – synthesis gas after combined conversion; VI – water; VII – dried synthesis gas; VIII – heated synthesis gas to methanol synthesis; IXa, IXb, IXc – synthesis gas with methanol; X – synthesis gas for recycling; XI – methanol; XII – hydrogen-containing gas (exhaust gas.)

for the value α = 0.38. **Table 2** presents the composition of the syngas obtained at the stage of oxygen matrix conversion of methane and CO_2 (**Figure 6**).

The catalytic reforming unit provides two processes, one of which is the steam methane conversion (stream Ib) and the subsequent steam conversion of CO obtained at the matrix stage (stream III) and during the catalytic steam methane conversion. The temperature linear gradient of the reaction in the reforming unit is set from 900°C at the beginning to 500°C at the end of the reactor. Pressure in the reformer is 10 bar. When calculating the catalytic steam conversion in the Aspen HYSYS program, we used the Gibbs reactor with the option "calculation of phase equilibrium and chemical equilibrium".

As in [17], the consumption of natural gas (methane) supplied at the stage of exothermic matrix and endothermic steam conversion was selected in such a way as to ensure the autothermal mode of the entire process when calculating the energy balance, considering the heat generated at the stage of methanol synthesis. It was also taken into account that, for efficient methanol synthesis and maximum COx conversion, it is desirable that the syngas functional $f = ([H_2] - [CO_2])/([CO] + [CO_2])$ be 2.0–2.2 [19].

Table 3 shows the main chemical reactions that take place in the catalytic reactor of methanol production, which are the hydrogenation of CO (Eq. (1)); the reverse water-gas shift reaction (Eq. (2)); and the hydrogenation of CO_2 (Eq. (3)) [19, 20]. Kinetics of these reactions are also presented in [21, 22]. It has been shown [23, 24] that the content of higher alcohols and ethers in the resulting methanol is low, less than 2 wt. %; therefore, their formation was neglected in this calculations.

Table 4 shows the parameters that we used to calculate the performance of the successive stages of the proposed method for methanol producing from methane and CO_2.

Component	CH_4	O_2	H_2	CO	CO_2	H_2O
Stream IV, vol. %	5.1	0	46.4	30.3	13.2	5.0

Table 2.
*Composition of syngas produced at the stage of oxygen matrix conversion of methane with CO_2 (**Figure 6**).*

No.	Chemical reaction	Reaction kinetics
1	$CO + 2H_2 \leftrightarrow CH_3OH$	$r_1 = \dfrac{k_1 K_{CO}\left(p_{CO} P_{H2}^{1.5} - p_{CH3OH} / \left(P_{H2}^{0.5} K_{p1}^{0}\right)\right)}{\left(1 + K_{CO} p_{CO} + K_{CO2} p_{CO2}\right)\left(p_{H2}^{0.5} + \left(K_{H2O}/K_{H2}\right)^{0.5} p_{H2O}\right)} \rho$
2	$CO_2 + H_2 \leftrightarrow CO + H_2O$	$r_2 = \dfrac{k_2 K_{CO2}\left(p_{CO2} p_{H2} - p_{H2O} p_{CO} / \left(K_{p2}^{0}\right)\right)}{\left(1 + K_{CO} p_{CO} + K_{CO2} p_{CO2}\right)\left(p_{H2}^{0.5} + \left(K_{H2O}/K_{H2}\right)^{0.5} p_{H2O}\right)} \rho$
3	$CO_2 + 3H_2 \leftrightarrow CH_3OH + H_2O$	$r_3 = \dfrac{k_3 K_{CO2}\left(p_{CO2} p_{H2}^{1.5} - p_{CH3OH} p_{H2O} / \left(P_{H2}^{1.5} K_{p3}^{0}\right)\right)}{\left(1 + K_{CO} p_{CO} + K_{CO2} p_{CO2}\right)\left(p_{H2}^{0.5} + \left(K_{H2O}/K_{H2}\right)^{0.5} p_{H2O}\right)} \rho$

r_i – reaction i (1–3) reaction rate [mol/(m^3s)]; k_i – reaction rate kinetic constants i (1–3) of methanol synthesis; p – partial pressure of components (bar); K – adsorption equilibrium constant of components (1/bar); ρ – gas species density [kg m^{-3}].

Table 3.
The main chemical reactions taking place in catalytic reactor of methanol synthesis and their kinetic parameters.

Parameter	Matrix conversion stage (combined oxidative-carbon dioxide conversion of methane)	Steam conversion of methane and CO stage	Methanol synthesis stage
Temperature, °C	1100–1200	900–500	250
Pressure, bar	10	10	50
C/steam mole ratio	—	1:3.0	—
Circulation	—	—	1:5

Table 4.
Initial data for calculating the process of methanol synthesis from methane and CO_2.

Component	Matrix conversion (streams Ia + II)		Steam conversion (stream Ib)		Dry syngas (stream VII)		Methanol (stream XI)	Exhaust gas (stream XII)	
	V, m^3/h	V, kg/h	V, m^3/h	V, kg/h	%vol	V, kg/h	V, kg/h	%vol	V, kg/h
CH_4	1311	861	2800	1840	0.1	16.6		6.6	17
O_2					0	0		0	0
H_2					58.2	1075		86.1	27
CO					15.6	4032		1.9	8.3
CO_2	328	644			8.5	3464		5.4	37.3
H_2O							1417		
CH_3OH							6479		

Table 5.
Material balance of the methanol production based on matrix conversion of natural gas with CO_2 addition.

When calculating the flow-circulation stage of methanol synthesis, the conversion of CO was set to 25% per pass [25]. The resulting material balance is presented in **Table 5**.

According to calculations, the functional of the resulting syngas is 2.1. This is the optimal value for methanol synthesis, which makes it possible to radically reduce the content of carbon oxides in the gas after the methanol production stage compared to their content in the initially produced hydrogen-containing gas [19]. The methanol concentration in the resulting raw methanol is more than 82 wt. %.

According to the preliminary material balance of the process, with a complete flow of incoming methane of 3111 m^3/h (2701 kg/h), the yield of methanol will be 6479 kg/h, when the COx output is less than 50 kg/h. Thus, the proposed method for producing methanol makes it possible to recycle up to 0.23 kg of CO_2 per 1 kg of methane.

4. Conclusions

The presented material shows the real possibility of utilizing a significant amount of CO_2 in the production of demanded petrochemical products from methane. The real possibility of organizing gas chemical processes with almost zero CO_2 emissions has also been demonstrated. This is the most realistic and economically feasible approach to reducing the emissions of this major greenhouse gas in technological processes.

Conflict of interest

The authors declare no conflict of interest.

Author details

Vladimir Arutyunov[1,2], Aleksey Nikitin[1,2]*, Valeriy Savchenko[2] and Igor Sedov[2]

1 N.N. Semenov Federal Research Center for Chemical Physics, Russian Academy of Sciences, Moscow, Russia

2 Federal Research Center of Problems of Chemical Physics and Medicinal Chemistry, Russian Academy of Sciences, Chernogolovka, Moscow Area, Russia

*Address all correspondence to: ni_kit_in@rambler.ru

References

[1] Saunois M, Martinez A, Poulter B, Zhang Z, Raymond PA, Regnier P, et al. Global methane budget 2000-2020. Earth System Science Data. 2025;**17**(5):1873-1958. DOI: 10.5194/essd-17-1873-2025

[2] Al-Enazi A, Okonkwo EC, Bicer Y, Al-Ansari T. A review of cleaner alternative fuels for maritime transportation. Energy Reports. 2021;**7**:1962-1985. DOI: 10.1016/j.egyr.2021.03.036

[3] Umeda M, Niitsuma Y, Horikawa T, Matsuda S, Osawa M. Electrochemical reduction of CO_2 to methane on platinum catalysts without overpotentials: Strategies for improving conversion efficiency. ACS Applied Energy Materials. 2020;**3**(1):1119-1127. DOI: 10.1021/acsaem.9b02178

[4] Cored J, García-Ortiz A, Iborra S, Climent MJ, Liu L, Chuang C-H, et al. Hydrothermal synthesis of ruthenium nanoparticles with a metallic core and a ruthenium carbide shell for low-temperature activation of CO_2 to methane. Journal of the American Chemical Society. 2019;**141**(49):19304-19311. DOI: 10.1021/jacs.9b07088

[5] Rihko-Struckmann LK, Peschel A, Hanke-Rauschenbach R, Sundmacher K. Assessment of methanol synthesis utilizing exhaust CO_2 for chemical storage of electrical energy. Industrial & Engineering Chemistry Research. 2010;**49**:11073-11078. DOI: 10.1021/ie100508w

[6] Arutyunov VS, Savchenko VI, Sedov IV, Nikitin AV. Non-catalytic gas phase oxidation of hydrocarbons. Eurasian Chemico-Technological Journal. 2022;**24**(1):13-20. DOI: 10.18321/ectj1144

[7] Guéret C, Daroux M, Billaud F. Methane pyrolysis: Thermodynamics. Chemical Engineering Science. 1997;**52**(5):815-827. DOI: 10.1016/S0009-2509(96)00444-7

[8] Wheatley J. Methane and the warming blame game. Methane. 2025;**4**(3):20. DOI: 10.3390/methane4030020

[9] Aaron D, Tsouris C. Separation of CO_2 from flue gas: A review. Separation Science and Technology. 2005;**40**(1-3):321-348. DOI: 10.1081/SS-200042244

[10] Brunetti A, Scura F, Barbieri G, Drioli E. Membrane technologies for CO_2 separation. Journal of Membrane Science. 2010;**359**(1-2):115-125. DOI: 10.1016/j.memsci.2009.11.040

[11] Nikitin AV, Ozersky AV, Savchenko VI, Sedov IV, Shmelev VM, Arutyunov VS. Matrix conversion of natural gas to syngas: The main parameters of the process and possible applications. Chemical Engineering Journal. 2019;**377**:129883. DOI: 10.1016/j.cej.2019.01.162

[12] Zhdanok S, Kennedy LA, Koester G. Superadiabatic combustion of methane air mixtures under filtration in a packed bed. Combustion and Flame. 1995;**100**(1-2):221-231. DOI: 10.1016/0010-2180(94)00064-Y

[13] Dorofeenko SO, Polianczyk EV. Conversion of hydrocarbon gases to synthesis gas in a reversed-flow filtration combustion reactor. Chemical Engineering Journal. 2016;**292**:183-189. DOI: 10.1016/j.cej.2016.02.013

[14] Savchenko VI, Zimin YS, Nikitin AV, Sedov IV, Arutyunov VS. Utilization

of CO_2 in non-catalytic dry reforming of C1–C4 hydrocarbons. Journal of CO_2 Utilization. 2021;**47**:101490. DOI: 10.1016/j.jcou.2021.101490

[15] Savchenko VI, Nikitin AV, Sedov IV, Ozerskii AV, Arutyunov VS. The role of homogeneous steam reforming of acetylene in the partial oxidation of methane to syngas in matrix type converters. Chemical Engineering Science. 2019;**207**:744-751. DOI: 10.1016/j.ces.2019.07.012

[16] Carapellucci R, Giordano L. Steam, dry and autothermal methane reforming for hydrogen production: A thermodynamic equilibrium analysis. Journal of Power Sources. 2020;**469**:228391. DOI: 10.1016/j.jpowsour.2020.228391

[17] Arutyunov VS, Nikitin AV, Savchenko VI, Sedov IV. Production of hydrogen and methanol without CO_2 emission based on matrix conversion of natural gas. Combined Doklady Chemistry. 2023;**513**:361-366. DOI: 10.1134/S0012500823601018

[18] Peng DY, Robinson DB. Industrial & engineering chemistry fundamentals. A New Two-Constant Equation of State. 1976;**15**(1):59-64

[19] Afanas'ev SV, Sadovnikov AA, Gartman VG, Obysov AV, Dul'nev AV. In: Afanas'ev SV, editor. Promyshlennyi Kataliz v Gazokhimii (Industrial Catalysis in Gas Chemistry). Chapter 5. Samara: Izd. SNTs RAN; 2018

[20] Graaf GH, Stamhuis EJ, Beenackers CM. Kinetics of low-pressure methanol synthesis. Chemical Engineering Science. 1988;**43**:3185-3195

[21] Hartig F, Keil FJ. Large-scale spherical fixed bed reactors: Modeling and optimization. Industrial & Engineering Chemistry Research. 1993;**32**:424-437

[22] Fu J, Majid MS, Altalbawy FM, Hussein RM, Waleed I, Mohammed IM, et al. Process simulation of methanol production via carbon dioxide hydrogenation. Case Studies in Thermal Engineering. 2024;**54**:103975. DOI: 10.1016/j.csite.2024.103975

[23] Narochnyi GB, Savost'yanov AP, Zubkov IN, Dulnev AV, Yakovenko RE. Investigation of the zinc-copper catalyst NIAP-06-06 for steam conversion of carbon monoxide in the synthesis of methanol. Kataliz v promyshlennosti (In Russian.). 2021;**21**(6):406-412. DOI:10.18412/1816-0387-2021-6-406-412

[24] Vanden Bussche KM, Froment GFA. Steady-state kinetic model for methanol synthesis and the water gas shift reaction on a commercial $Cu/ZnO/Al_2O_3$. Journal of Catalysis. 1996;**161**:1-10

[25] Zagashvili YUV. An indicator of the degree of carbon conversion during the conversion of synthesis gas to methanol. Oil & Gas Chemistry. 2025;**№ 1-2**:48-52. DOI:10.24412/2310-8266-2025-1-2-48-52

Chapter 3

Histopathological Consequences of Methane and Associated Air Pollutants: Linking Oxidative Stress to Organ Toxicity and Carcinogenesis

Akinpelu Moronkeji and Temidayo Daniel Adeniyi

Abstract

Anthropogenic human activities like mining, fertilizer production as well as various industrial processes have been linked to the rising prevalence of non-communicable diseases, including cardiovascular disease, diabetes, lung cancer and leukemia. These activities significantly contribute to the increasing levels of methane and associated air pollutants, which play a critical role in cellular dysfunction, toxicity, and cancer development. Air pollution remains a major global health challenge, accounting for substantial morbidity and mortality worldwide. Methane and its associated air toxicants have been shown to induce oxidative stress, genotoxicity, hepatorenal toxicity and haematotoxicity, leading to histoarchitectural alterations in vital organs such as the brain, lungs, heart, liver and kidneys. These pollutants disrupt cellular homeostasis and promote inflammation, DNA damage and tumourigenesis. This chapter explores the mechanisms underlying methane-associated oxidative stress and histological changes in mammalian cells, highlighting their oncogenic potential and the broader implications of air pollution on public health.

Keywords: cancer, pollution, toxicity, oxidative stress, histopathology

1. Introduction

Air pollution is a significant environmental and public health issue characterized by the presence of harmful or excessive levels of substances in breathable air [1, 2]. These pollutants are either naturally occurring or anthropogenic and can exist in the form of gases, particulate matter, or biological [3–5]. Particulate matter (PM), nitrogen oxides (NO_x), sulfur dioxide (SO_2), carbon monoxide (CO), ozone (O_3), volatile organic compounds (VOCs) and heavy metals such as lead (Pb) and mercury (Hg) are important pollutants associated with various diseases [6–8]. These pollutants originate from both natural events such as volcanic eruptions and wildfires, as well as human activities like industrial operations and transportation [9–11].

Air pollution significantly increases the risk of respiratory conditions, including asthma and chronic obstructive pulmonary disease (COPD), with long-term exposure to pollutants such as SO_2, NO_2 and PM impairing lung function, elevating respiratory disorder prevalence while also driving up hospital admission levels [12–14]. Fine particulate matter ($PM_{2.5}$) remains one of the most dangerous airborne contaminants that can penetrate deep into the lungs and traverse into the bloodstream, adversely affecting the cardiovascular system [15]. Exposure to $PM_{2.5}$ has been linked to elevated blood pressure, thrombosis, insulin resistance, vascular injury, and accelerated atherosclerosis [16–19]. These effects are largely mediated through mechanisms involving oxidative stress, inflammation and autonomic nervous system dysregulation, ultimately increasing the risk of heart failure, ischemic heart disease and stroke [20]. Additionally, emerging evidence also supports a strong association between air pollution and neurological disorders. Specifically, long-term $PM_{2.5}$ exposure has been correlated with increased risks of stroke, dementia, Alzheimer's disease, Parkinson's disease and autism spectrum [21–24].

Globally, air pollution remains a critical public health problem, contributing to an estimated 5.2 million premature deaths annually, with 7.3 billion people exposed to unsafe $PM_{2.5}$ levels. Low- and middle-income countries, where nearly 80% of the affected population resides, bear the greatest burden due to the widespread use of polluting technologies [25, 26]. This review provides a mechanistic histopathological insight into the health impacts of methane and associated air pollutants, highlighting their disproportionate burden on vulnerable populations and the urgent need for action to advance global health equity and achieve sustainable development goals.

1.1 Methane as an air pollutant: Sources, production, and global significance

Methane (CH_4) is a potent greenhouse gas with a high global warming potential, contributing approximately 30% to global warming [27]. It originates from both natural and anthropogenic sources, including wet forest soils, fuel combustion, agriculture, and landfills [28, 29]. The production of methane primarily occurs in anaerobic environments through microbial decomposition of organic matter. In deep aquifers, such as those in accretionary prisms, syntrophic interactions between fermentative bacteria and methanogenic archaea are critical for methane generation [30]. The main methanogenic pathways, aceticlastic and hydrogenotrophic, vary in dominance depending on factors like temperature and substrate availability [31].

Methane's significant impact on climate change stems from its higher global warming potential compared to CO_2 over short timeframes, contributing about 15–20% to the enhanced greenhouse effect [32, 33]. Reducing methane emissions offers a "win-win" benefit by mitigating climate change and decreasing air pollution, as methane is a precursor to ground-level ozone [32]. The Global Methane Pledge aims to cut emissions by 30% by 2030, which could yield greater near-term cooling than comparable CO_2 reductions [34]. Although advances in methane detection and mitigation, especially within the oil and gas sectors, have progressed, widespread implementation remains limited [35].

1.2 Environmental interactions: How methane contributes to air pollution dynamics

Methane plays a complex role in air pollution by contributing to the formation of secondary pollutants and affecting climate, ecosystems, and human health. As a precursor to tropospheric ozone, methane indirectly affects the respiratory and cardiovascular systems [6]. Through multiphase atmospheric chemical reactions, methane contributes to the formation of secondary organic aerosols (SOAs) and other harmful compounds [36, 37]. Despite its significance, methane's governance under climate frameworks primarily considers its CO_2-equivalent impact over 100 years,

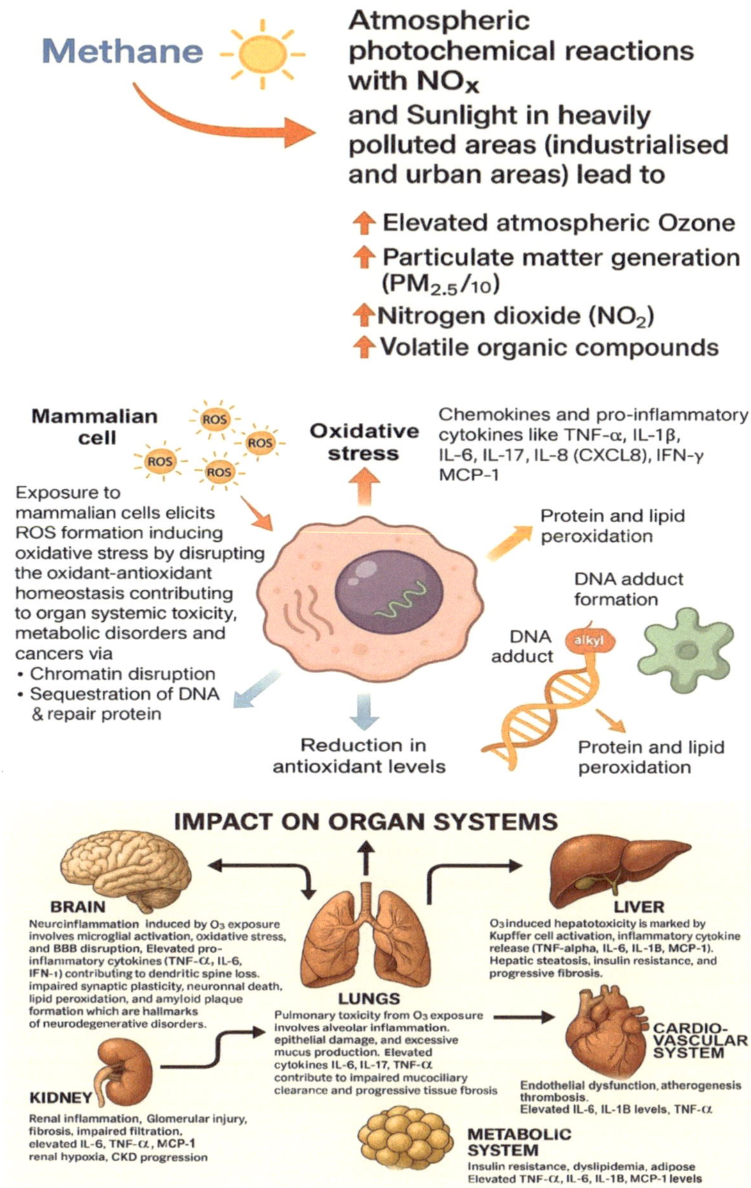

Figure 1.
Interaction of methane and associated air pollutants on the mammalian system. Methane photochemically reacts with sunlight and NO_x in polluted industrial and urban areas leading to the formation of SOAs, O_3, $PM_{2.5}/_{10}$, and NO_2. In mammalian cells, these pollutants generate ROS, disrupting redox balance and eliciting oxidative stress, inflammation, DNA damage, and chromatin alterations. The resulting expression of chemokines (TNF-α, IL-1β, IL-6, IL-17, IL-8, IFN-γ, MCP-1), alongside protein and lipid peroxidation with reduced antioxidant.

overlooking its potent near-term climate effects and role in air pollution [32]. Tropospheric ozone forms when methane undergoes photochemical reactions with sunlight and nitrogen oxides (NO_x), especially in urban and industrial regions where NO_x emissions are high [38]. Elevated methane levels increase ground-level ozone concentrations, contributing to respiratory illnesses, agricultural damage, and ecosystem disruption. Globally, ozone-related respiratory deaths attributable to methane emissions are estimated at 760 per million metric tons, potentially doubling methane's social cost [39]. Although methane is not a direct source of particulate matter, its atmospheric oxidation produces volatile organic compounds like formaldehyde, which contribute to the formation of secondary organic aerosols and $PM_{2.5}$, thereby indirectly increasing fine particulate matter concentrations [40, 41]. Methane's atmospheric lifetime and climate impact are strongly influenced by hydroxyl radicals (OH), the primary atmospheric oxidant. Since 1980, global OH levels have risen by about 9%, driven by changes in anthropogenic emissions [42, 43]. Under carbon neutrality scenarios, continued OH increases could shorten methane's lifetime and reduce its radiative forcing, whereas declining OH levels in unmitigated scenarios would intensify methane's climate impact. This interaction introduces significant uncertainty in global methane budget estimates comparable to uncertainties in emission inventories (**Figure 1**) [43, 44].

2. Histopathological effects of air pollution on organ system

2.1 Neurotoxicity and neuroinflammation associated with air pollutants

The blood-brain barrier (BBB) is a specialized system composed of endothelial cells, pericytes, astrocytes, and other components that regulate exchange between the bloodstream and brain tissue, protecting the central nervous system (CNS) from toxic agents and maintaining neural homeostasis [45, 46]. The BBB dysfunction is characterized by disruption of tight junctions (TJs) between endothelial cells and is implicated in aging and neurological disorders such as stroke, Alzheimer's disease, and multiple sclerosis [47]. The reduced expression of TJ proteins like claudin-5 has been linked to psychiatric conditions including depression and schizophrenia [48]. Similarly, neonatal hypoxic-ischemic encephalopathy and brain injury result in TJ degradation, increased BBB permeability, and elevated inflammatory cytokines and oxidative stress [49].

Air pollutants such as PM and O_3 compromise the blood-brain barrier, facilitating the influx of neurotoxic substances and inflammatory mediators into the CNS. This triggers microglial activation, leading to the production of reactive oxygen species and proinflammatory cytokines, resulting in oxidative stress and neuronal damage. This chronic neuroinflammation contributes to the development of neurodegenerative disorders such as Alzheimer's disease, Parkinson's disease, multiple sclerosis, stroke, and cognitive decline [50]. Growing evidence also indicates that air pollution adversely impacts CNS development and aging, disrupting neurodevelopmental trajectories and accelerating neurodegeneration [51].

2.1.1 Potential links to neurodegenerative diseases

Methane's impact on neurodegenerative diseases is likely indirect, primarily through its role in forming other harmful air pollutants rather than direct neurotoxicity. Notably, tropospheric ozone (O_3), which is produced from methane, is strongly linked to neurodegenerative pathology. Chronic low-dose exposure to O_3 induces oxidative stress, leading to dendritic spine loss, impaired synaptic plasticity, and neuronal death [52]. O_3 can penetrate the blood-brain barrier (BBB), causing oxidative damage, mitochondrial dysfunction, and lipid peroxidation in neuron processes central to dementia pathology, especially Alzheimer's disease [53]. O_3 exposure also exacerbates oxidative stress, disrupts metabolic pathways, and promotes amyloid plaque accumulation, contributing to Alzheimer-like pathology [10]. Chronic exposure to air pollutants further promotes neuropathological markers of neurodegeneration, especially in genetically susceptible individuals. The hippocampus is particularly vulnerable due to its limited antioxidant defenses and direct anatomical connectivity with the olfactory epithelium, a potential route for pollutants to enter the brain [54].

2.2 Synergistic effects of methane in worsening pulmonary conditions

Although traditionally considered inert, emerging evidence indicates that methane exhibits bioactive properties capable of modulating oxidative stress responses and mitochondrial function upon inhalation [55]. While not directly toxic at typical environmental concentrations, methane plays a critical role in atmospheric chemistry by contributing to the formation of ground-level ozone (O_3), a well-established respiratory irritant [32]. Methane-driven ozone, formed when methane and nitrogen oxides (NO_x) react under sunlight, exacerbates respiratory conditions such as asthma and chronic obstructive pulmonary disease (COPD) by inducing airway inflammation and dysfunction [56, 57]. This methane-derived ozone, combined with particulate matter, synergistically worsens respiratory health by impairing mucociliary clearance and promoting lung tissue fibrosis [58, 59]. Additionally, non-methane hydrocarbons (NMHCs) contribute alongside methane to air pollution's respiratory toxicity, with tropospheric ozone recognized as a critical mediator of pulmonary irritation and inflammation [32].

2.3 Impact on endothelial function and vascular inflammation

Environmental stressors such as air pollution substantially contribute to the global burden of cardiovascular diseases, with oxidative stress being a central pathway mediating these effects [6, 11, 60, 61]. Endothelial dysfunction, typically marked by reduced nitric oxide (NO) bioavailability, is pivotal in cardiovascular pathologies [62]. Pollutants, including heavy metals and metal oxide nanoparticles, induce oxidative stress, endothelial apoptosis, and activation, disrupting vascular tone, triggering inflammation, and altering platelet activity [63, 64]. Activated endothelial cells upregulate adhesion molecules, such as ICAM-1 and VCAM-1, facilitating leukocyte adhesion and migration, which propagate vascular inflammation [65]. These effects are mediated by proinflammatory cytokines, oxidized lipoproteins, and nuclear factor-κB [NF-Kb] signaling pathways [66]. Epigenetic mechanisms including DNA methylation and histone modifications regulate adhesion molecule expression, further complicating endothelial activation [67]. Such activation fosters a pro-thrombotic state that accelerates atherosclerosis progression, making ICAM-1 and VCAM-1 promising biomarkers and therapeutic targets in cardiovascular disease [65, 66].

2.4 Evidence of pollutant-induced kidney damage

Air pollution, even at low to moderate exposure levels, can induce chronic inflammation and oxidative stress, thereby contributing to the onset and progression of kidney [68–71]. Methane itself is not directly nephrotoxic but may indirectly contribute to renal injury through mechanisms involving renal hypoxia, oxidative stress, and metabolic disturbances [72]. Renal hypoxia is a key driver in chronic kidney disease (CKD) progression. Oxidative stress, arising from an imbalance between ROS and antioxidant defenses, plays a central role in kidney injury [73]. The kidney's high mitochondrial density, responsible for about 90% of cellular ROS production, makes it particularly susceptible to oxidative damage [74]. Mitochondrial dysfunction disrupts metabolism and dynamics, activating redox-sensitive pathways that promote inflammation and apoptosis, accelerating CKD [75, 76]. Long-term ozone exposure is associated with reduced estimated glomerular filtration rate (eGFR) in middle-aged and elderly adults, with acute and chronic exposures linked to kidney function decline, potentially mediated by hyperglycaemia and insulin resistance [77, 78]. Proinflammatory cytokines, particularly TNF-α, contribute to renal damage through TNF receptor signaling pathways [79]. Cellular senescence and the senescence-associated secretory phenotype (SASP) exacerbate chronic inflammation in CKD, further driving disease progression [80].

3. Oxidative stress and cellular damage mechanism

3.1 Mechanism of oxidative stress: Role of ROS and RNS

Oxidative stress results from an imbalance between the generation of ROS and RNS and the body's antioxidant defenses, causing damage to proteins, lipids and DNA [11, 81, 82]. ROS are produced endogenously primarily by mitochondria and NADPH oxidases and exogenously from sources such as UV radiation and environmental pollutants [83]. Both ROS such as superoxide anion, hydrogen peroxide, hydroxyl radical and RNS like nitric oxide, and peroxynitrite are critical in cellular signaling and stress responses [84, 85]. While physiological levels of ROS/RNS support normal biochemical pathways, their excess leads to oxidative damage [86].

Mitochondria maintain redox balance through enzymes that degrade hydrogen peroxide and negative feedback mechanisms [87]. Excessive ROS/RNS generation causes oxidative modifications like lipid peroxidation and protein/DNA damage, contributing to diseases such as cancer, cardiovascular, and neurodegenerative disorders [88, 89]. The antioxidant defense system includes enzymatic components like superoxide dismutase (SOD), catalase (CAT), glutathione peroxidase (GPx), and non-enzymatic antioxidants such as vitamins C and E, which serve as defenses and neutralize ROS to maintain the cellular redox homeostasis [90–92]. However, chronic oxidative stress can overwhelm these systems, leading to pathological conditions like diabetes, cancer, and neurodegeneration [93].

3.2 Biomolecular damage and associated oxidative biomarkers: Lipid peroxidation, protein oxidation, and DNA damage

Lipid peroxidation, protein oxidation, and DNA damage are interrelated processes that compromise cellular integrity and contribute to aging and the pathogenesis of various diseases. Among these, lipid peroxidation is a hallmark of ferroptosis, which is a regulated form of cell death marked by the oxidative degradation of polyunsaturated fatty acids in cellular membranes, leading to the accumulation of lipid hydroperoxides (LOOH) [88, 94]. Oxidative stress-induced lipid peroxidation generates reactive aldehydes, such as 4-hydroxynonenal (4-HNE) and malondialdehyde (MDA), with 4-HNE forming adducts with proteins and DNA that disrupt cellular function and signaling. These adducts are key biomarkers of oxidative damage in conditions like neurodegenerative diseases and cancer [95, 96]. Mitochondria are both sources and targets of 4-HNE, with adduct formation impairing respiratory complexes and promoting apoptosis [97]. ELISA assays for 4-HNE-protein adducts have demonstrated elevated levels in patient samples, underscoring their clinical relevance [98]. Protein oxidation results from ROS/RNS-mediated modifications that alter protein structure and function [99]. Advanced oxidation protein products (AOPPs) and 3-nitrotyrosine are widely recognized biomarkers reflecting oxidative protein damage [100]. These oxidative changes can disrupt cellular homeostasis through protein dysfunction and aggregation. However, some oxidative modifications serve physiological roles in redox signaling and metabolism [101]. 8-Hydroxy-2′-deoxyguanosine (8-OHdG) is a primary biomarker of oxidative DNA damage, formed via ROS oxidation of guanine bases [102]. Elevated 8-OHdG levels correlate with cancer, neurodegenerative diseases, and aging [103]. High-performance liquid chromatography with electrochemical detection (HPLC-ECD) is commonly employed for sensitive and selective quantification [102]. Urinary 8-OHdG provides a non-invasive measure of oxidative stress, with smoking linked to higher levels, while no gender differences have been noted [104].

Glutathione peroxidase 4 (GPx4) plays an integral role in mitigating lipid peroxidation by reducing LOOH; however, ferroptosis is triggered when this antioxidant defense is compromised, typically due to iron overload and glutathione depletion [94, 105, 106]. The accumulation of oxidized lipids disrupts organelle function and promotes non-apoptotic cell death [106].

The oxidation of protein involves ROS/RNS-mediated modifications of amino acids such as tyrosine, methionine, cysteine, and tryptophan, leading to carbonylation, nitration, structural changes, and functional loss [101, 107]. Some oxidative modifications are reversible and modulate protein function, whereas others cause aggregation and proteotoxicity [101]. Redox-sensitive proteases maintain homeostasis by mediating pro-survival or cell death pathways in response to oxidative stress [108]. Oxidative DNA damage contributes to genomic instability and neurodegenerative disease progression. ROS alter DNA bases, causing mutations and impairing genome function, sometimes serving as epigenetic marks [109]. Age-related decline in DNA damage response exacerbates this damage, playing a role in disorders like Alzheimer's and Parkinson's diseases [110]. Protein aggregates can disrupt chromatin and sequester DNA repair proteins, increasing genomic instability [111]. ROS-induced lipid peroxidation generates reactive carbonyl species (RCS) that further damage DNA and proteins, promoting misfolding and aggregation characteristic of neurodegeneration [89, 112]. The accumulation of oxidized biomolecules overwhelms cellular clearance mechanisms, leading to organelle dysfunction and cell death.

Chronic oxidative stress is implicated in cancer, cardiovascular diseases, and neurodegeneration, while antioxidants mitigate damage by scavenging free radicals and chelating metals [106, 113]. Methane-associated air pollution may exacerbate oxidative stress indirectly by generating secondary pollutants that increase systemic ROS levels, promoting DNA damage, mitochondrial dysfunction, and redox imbalance, which are key factors that heighten the risks of neurodegenerative and chronic diseases.

3.3 Antioxidant enzyme biomarkers

Key antioxidant enzymes like superoxide dismutase (SOD), catalase (CAT), and glutathione peroxidase (GPx) constitute the frontline defense against ROS-mediated damage [114, 115]. Altered activities of these enzymes serve as biomarkers in diseases such as cancer and depression, aiding diagnosis and prognosis [116, 117]. The enzymatic antioxidant system, including thioredoxin and paraoxonase, maintains redox homeostasis across pathological conditions like diabetes, autoimmune disorders, and neurodegeneration [118]. Reduced expression of antioxidant enzymes is associated with increased oxidative damage and worsened disease outcomes. For example, lower GPX1, SELENOP, and CAT expression correlates with poor survival in hepatocellular carcinoma [119, 120]. Conversely, in multiple sclerosis, the upregulation of antioxidant enzymes in microglia/macrophages may reflect protective responses to oxidative injury [121]. Chronic oxidative stress disrupts the balance between oxidants and antioxidants, contributing to the progression of cancer, atherosclerosis, and neurodegenerative diseases [122].

3.4 Histopathological techniques in oxidative stress assessment

Histopathological methods are essential for detecting and quantifying oxidative stress biomarkers within tissues. Immunohistochemistry (IHC) is widely employed to localize oxidative stress markers at the cellular level, particularly in neurodegenerative disease models and postmortem samples [123]. Quantitative assessment is often achieved via Western blotting and enzyme-linked immunosorbent assays (ELISA), with ELISA commonly used for clinical sample analysis and oxidative stress evaluation [124]. Fluorescence-based imaging techniques provide enhanced visualization of reactive oxygen species (ROS) and oxidative damage at subcellular resolution [125]. Emerging analytical approaches such as DNA damage sequencing, chemical isotope labelling and mass spectrometry imaging offer advanced capabilities for the comprehensive analysis of oxidative stress biomarkers (**Table 1**) [126].

S/N	Organ system	Pathological features	Histopathological techniques	Output
1.	Brain (neurotoxicity)	Neuronal degeneration, pyknosis	Hematoxylin and eosin (H&E)	Necrotic/degenerative neurons.
	Demyelination	Demyelination	Luxol fast blue (LFB) for selective myelin staining, with loss of myelin indicating demyelination	Demyelination
	Reactive gliosis	Astrocytic proliferation (gliosis)	Glial fibrillary acidic protein (GFAP) immunostaining Astrocytic proliferative marker (gliosis marker).	Reactive gliosis
	Microglial activation	Activated microglia	Immunostaining (IHC) techniques for highlighting activated microglia using the Ionized calcium-binding adaptor molecule 1 (Iba 1).	Microglial activation
	Oxidative stress	DNA/RNA oxidative damage.	DNA/RNA oxidative damage markers using IHC. 8-Hydroxy-2′-deoxyguanosine (8-OHdG).	Oxidative stress
2	Liver (hepatotoxicity)	Hepatic steatosis	Frozen section using the Oil Red O staining protocol.	Identifies neutral lipids and fatty changes
		Ballooning, necrosis	H&E.	Detects hepatocyte damage and apoptosis
		Inflammation	IHC for CD68	Kupffer cell infiltration

S/N	Organ system	Pathological features	Histopathological techniques	Output
		Fibrosis	Sirius Red, Masson's trichrome.	Collagen fiber deposition associated with chronic exposure to toxicants.
3.	Kidneys (nephrotoxicity)	Tubular necrosis	H&E	General morphology and necrosis
		Basement membrane thickening	Periodic acid-schiff (PAS), Jones silver stain	Glomerular and tubular basement membrane changes
		Interstitial fibrosis	Masson's trichrome	Highlights fibrotic changes and collagen.
		Oxidative injury	IHC for 8-OHdG and nuclear factor erythroid 2-related factor 2 (Nrf2).	DNA oxidative damage and antioxidant response marker
4	Lungs (pulmonary toxicity)	Fibrosis	Masson's trichrome or Van Gieson	Collagen deposition and fibrotic thickening
		Inflammatory infiltration	H&E, CD45 IHC	Identification of immune cells (e.g., lymphocytes, neutrophils)
		Mucus production	Periodic acid-schiff [PAS]	Highlights goblet cell metaplasia and mucus
		Epithelial hyperplasia	H&E, Ki-67 IHC	General cell proliferation and hyperplasia marker
5	General (multi-organ)	Lipid peroxidation	Immunostaining for 4-hydroxynonenal (4-HNE), malondialdehyde (MDA)	Detects products of lipid peroxidation
		DNA damage	Immunostaining of γ-phosphorylated form of the histone H2AX (γ-H2AX), Comet Assay.	DNA double-strand breaks and genotoxic stress
		Inflammation	CD3 (T cells), CD68 (macrophages)	Immune cell profiling
		Hypoxia	Immunostaining for hypoxia-inducible factor 1 alpha (HIF-1α).	Marker of cellular response to hypoxia

Table 1.
Histopathological features and diagnostic techniques in pollution-exposed organ system.

4. Carcinogenic effects of methane and air pollutants

4.1 Mechanisms of pollutant-induced carcinogenesis

Carcinogenesis resulting from environmental pollutants involves intricate genetic and epigenetic alterations that disrupt cellular homeostasis. Epigenetic mechanisms such as DNA methylation, histone modifications, and regulation by non-coding RNAs are critical in lung cancer development and progression [127]. These epigenetic changes can be modulated by environmental exposures, transforming normal cells into malignant phenotypes. The exposome concept, encompassing all environmental exposures throughout life, interacts dynamically with individual genetic and epigenetic profiles to influence disease risk [128]. Particulate matter (PM) exposure induces oxidative stress, DNA damage, and genotoxicity, facilitating malignant transformation [129]. Both PM and polycyclic aromatic hydrocarbons (PAHs) contribute to carcinogenesis by forming DNA adducts and disrupting DNA repair pathways including base excision, nucleotide excision, and homologous recombination repair [130]. Cytochrome P450 enzymes bioactivate many carcinogenic chemicals found in pollutants, producing reactive intermediates that covalently bind to DNA, forming mutagenic adducts [131, 132]. The ROS generated by pollutants is a key mediator of oxidative stress, leading to DNA damage, genomic instability, and mutations associated with cancer and neurodegenerative diseases [7, 109]. ROS also induce lipid peroxidation, generating reactive carbonyl species that form adducts with DNA and proteins, disrupting critical cellular signaling pathways [89]. PM exacerbates oxidative damage by promoting DNA adduct formation, double-strand breaks, and impairing DNA repair mechanisms [130]. Chronic inflammation induced by environmental agents such as asbestos fosters a pro-tumourigenic microenvironment. This involves the release of proinflammatory cytokines, chemokines, growth factors, and ROS [133, 134]. ROS act as a critical link between inflammation and tumorigenesis by causing DNA damage and regulating cellular processes like proliferation, differentiation, and apoptosis [134]. The inflammatory milieu activates oncogenic signaling pathways including STAT3, NF-κB, and PI3K/Akt, stimulating the secretion of pro-angiogenic factors that support tumor vascularization and growth [135]. Oxidative stress also modulates epigenetic machinery, contributing to cancer progression and therapeutic resistance [136]. Epigenetic silencing of tumor suppressor genes or activation of oncogenes drives malignancy in various cancers [137].

4.2 Role of methane in potentiating the mutagenic effects of other pollutants

Emerging evidence highlights complex interactions between methane and other environmental pollutants in exacerbating adverse health effects [32, 110, 138]. Although methane is generally considered less reactive, it modulates oxidative stress responses and mitochondrial function, thereby potentiating the carcinogenic effects of more reactive contaminants [55, 139]. Methane contributes indirectly to carcinogenesis by participating in photochemical reactions with volatile organic compounds (VOCs) and PAHs, forming ground-level ozone, which is a potent oxidant that damages DNA and impairs cellular repair mechanisms [140]. These atmospheric oxidation processes can generate compounds capable of causing direct DNA damage without the need for metabolic activation [141]. While PAHs have been extensively studied for their genotoxicity, both the parent PAHs and their metabolites can exert non-genotoxic effects by activating diverse cellular pathways implicated in carcinogenesis [142]. The interplay between environmental pollutants, oxidative stress, and carcinogenesis is strongly linked to particulate matter exposure, which induces oxidative stress leading to genotoxicity and potential malignant transformation [11, 129]. Environmental mutagens collectively shape the mutational landscape in tissues, influencing cancer development [143]. Various mutagens induce DNA damage, triggering genetic and epigenetic alterations that may progress to cancer if DNA repair pathways fail [144, 145]. Interestingly, methane production occurs endogenously across all living organisms, including human cells, often as a response to oxidative stress [146, 147]. This intrinsic methane generation may influence cellular redox balance and contribute to the biological effects of environmental pollutants (**Table 2**).

Mechanisms	Consequences associated with air pollutants
Oxidative stress	Lipid peroxidation, DNA damage, mitochondrial dysfunction.
Inflammation	Cytokine cascade, neutrophilic infiltration, and chronic immune cell presence.
Hypoxia	Neuronal and myocardial injury, tubular necrosis.
Secondary pollutants (e.g., O_3 and NO_2)	Epithelial/endothelial damage, fibrosis, oedema.

Table 2.
Pathological drivers associated with environmental pollutants.

4.3 Histopathological evidence of cancer in the brain, liver, lungs, heart, and kidney

Glioblastomas (GBM) are aggressive primary brain tumors, exhibiting distinct histopathological features such as nuclear atypia, cellular pleomorphism, necrosis, and pseudopalisading necrosis [148]. These morphological abnormalities correspond to underlying molecular and cellular dysfunctions, including mutations, dysregulation of critical signaling pathways, and alterations in nuclear envelope proteins [149]. GBMs display profound genetic, transcriptomic, and cellular heterogeneity both within and between tumors, contributing to their malignancy and resistance to treatment. The tumor microenvironment is composed of various cell types, including tumor cells, immune infiltrates, and cancer stem cells that interact through complex signaling networks, promoting tumor growth and invasion [150].

Hepatocellular carcinoma (HCC) is the predominant primary malignancy of the liver, frequently associated with chronic liver disease and cirrhosis [151]. Histologically, HCC is characterized by atypical hepatocytes arranged in trabecular patterns, nuclear pleomorphism, and vascular invasion [152]. Although imaging modalities are primarily used for diagnosis, biopsy remains essential in atypical or ambiguous cases [153, 154]. Advances in genomics and transcriptomics have revealed recurrent genetic mutations and molecular subtypes correlated with histopathological variations, enhancing our understanding of HCC pathogenesis and prognostic stratification [155].

Lung cancers, predominantly adenocarcinomas and squamous cell carcinomas (SCC), exhibit distinct histological and molecular signatures. SCC is typified by keratinization and intercellular bridges [156]. Biomarkers such as Desmoglein 3 (DSG3) and Desmocollin 3 (DSC3) have been identified as key transcriptional features of SCC, with their expression levels correlating with patient prognosis [157–159]. Keratin 14 (KRT14) is also proposed as a biomarker distinguishing lung cancer subtypes and is associated with poorer outcomes in SCC patients [160]. These genes function in pathways involving the p63 transcription factor network and trail signaling, impacting tumor behavior and prognosis.

Cardiac angiosarcomas are rare but highly aggressive malignancies arising from endothelial cells lining cardiac blood vessels [161]. Histologically, they exhibit infiltrative growth with abnormal vascular channels and marked anaplasia [162]. These tumors often present at advanced stages, carrying a poor prognosis with a 5-year survival rate of 35% [163]. Diagnosis involves a multimodal approach, including echocardiography, CT, MRI, histopathology, and immunohistochemistry [161, 164]. Clinically, they may mimic other cardiac conditions and present symptoms such as cardiac tamponade, pleural effusion, and dyspnea [165, 166].

Clear cell renal cell carcinoma (ccRCC) is the most common subtype of kidney cancer, characterized histologically by nests of clear cells with abundant cytoplasm and a prominent vascular network [167]. ccRCC is frequently associated with the inactivation of the VHL gene. Single-cell RNA sequencing has uncovered substantial tumor microenvironment heterogeneity, including a vascular subpopulation linked to epithelial-mesenchymal transition, which may contribute to tumor progression and metastasis [168].

5. Conclusion

Although traditionally considered inert and non-toxic, methane undergoes atmospheric oxidation to generate volatile organic compounds (VOCs), which contribute to the formation of secondary organic aerosols, fine particulate matter, and ground-level ozone (O_3), a highly reactive pollutant. These pollutants interact with cellular structures and compromise multiple organ systems, notably the central nervous system, by disrupting the integrity of the blood-brain barrier. This interaction facilitates the uptake of neurotoxic substances, which in turn promotes neuroinflammation, dendritic spine loss, impaired synaptic plasticity, mitochondrial dysfunction, neuronal lipid peroxidation, and amyloid plaque formation which are hallmarks associated with neurodegenerative diseases. Furthermore, the reactive intermediates generated through these atmospheric chemical reactions elevate reactive oxygen species (ROS) levels, triggering pro-inflammatory cytokine release and driving oxidative modifications such as lipid peroxidation, protein oxidation, and DNA adduct formation. These molecular events act as precursors to a spectrum of chronic metabolic and degenerative conditions, including renal injury, chronic kidney disease, cardiovascular and respiratory disorders, and various forms of cancer. Collectively, these findings underscore the significant histopathological and systemic health risks associated with methane-derived air pollutants and emphasize the urgent need for integrated environmental and public health strategies to reduce exposure and disease burden.

Conflict of interest

The authors declare no conflict of interest.

Abbreviations

NMHCs	non-methane hydrocarbons
PM	particulate matter
ROS	reactive oxygen species
RNS	reactive Nitrogen Species
VOCs	volatile organic compounds
CH_4	methane
O_3	ozone
IHC	immunohistochemistry
VOCs	volatile organic compounds
PAHs	polycyclic aromatic hydrocarbons
AOPPs	advanced oxidation protein products

8-OhdG	8-hydroxy-2′-deoxyguanosine
HPLC-ECD	high-performance liquid chromatography with electrochemical detection
NO_x	nitrogen oxides
SO_2	sulfur dioxide
CO	carbon monoxide
COPD	chronic obstructive pulmonary disease
SOAs	secondary organic aerosols
OH	hydroxyl radicals
BBB	blood-brain barrier
CNS	central nervous system
TJs	tight junctions
CKD	chronic kidney disease

Author details

Akinpelu Moronkeji[1]* and Temidayo Daniel Adeniyi[2]

1 Department of Medical Laboratory Science, Faculty of Allied Health Sciences, University of Medical Sciences Ondo, Ondo City, Ondo State

2 Department of Medical Laboratory Science, Faculty of Basic Clinical Sciences, University of Ilorin, Ilorin, Kwara State, Nigeria

*Address all correspondence to: amoronkeji@unimed.edu.ng

References

[1] Lawin H, Fanou LA, Hinson AV, Stolbrink M, Houngbegnon P, Kedote NM, et al. Health risks associated with occupational exposure to ambient air pollution in commercial drivers: A systematic review. International Journal of Environmental Research and Public Health. 2018;**15**(9):2039. DOI: 10.3390/ijerph15092039

[2] Wei Y, Zhang J, Li Z, Gow A, Chung KF, Hu M, et al. Chronic exposure to air pollution particles increases the risk of obesity and metabolic syndrome: Findings from a natural experiment in Beijing. The FASEB Journal. 2016;**30**(6):2115-2122

[3] Orellano P, Reynoso J, Quaranta N, Bardach A, Ciapponi A. Short-term exposure to particulate matter (PM10 and PM2.5), nitrogen dioxide (NO_2), and ozone (O_3) and all-cause and cause-specific mortality: Systematic review and meta-analysis. Environment International. 2020;**142**(December 2019):105876. DOI: 10.1016/j.envint.2020.105876

[4] Traboulsi H, Guerrina N, Iu M, Maysinger D, Ariya P, Baglole CJ. Inhaled pollutants: The molecular scene behind respiratory and systemic diseases associated with ultrafine particulate matter. International Journal of Molecular Sciences. 2017;**18**(2):243. DOI: 10.3390/ijms18020243

[5] Moronkeji A, Adeniyi TD, Atere AD. Oxidative stress response to gasoline generator exhaust emission in adult male wistar rats. Environmental Analysis, Health and Toxicology. 2024;**39**(4):2-11

[6] Manisalidis I, Stavropoulou E, Stavropoulos A, Bezirtzoglou E. Environmental and health impacts of air pollution: A review. Frontiers in Public Health. 2020;**8**(February):1-13

[7] Adeniyi T, Moronkeji A, Oluwatunase G, Adekomi A, Adunmo G. Mercury-induced oxidative stress and neuropathologic changes in the prefrontal cortex of wistar rats. Nigerian Journal of Neuroscience. 2024;**15**(December):97-104

[8] Meo SA, Salih MA, Al-Hussain F, Alkhalifah JM, Meo AS, Akram A. Environmental pollutants PM2.5, PM10, carbon monoxide (CO), nitrogen dioxide (NO_2), sulfur dioxide (SO_2), and ozone (O_3) impair human cognitive functions. European Review for Medical and Pharmacological Sciences. 2024;**28**(2):789-796

[9] Shetty SS, Deepthi D, Harshitha S, Sonkusare S, Naik PB, Kumari NS, et al. Environmental pollutants and their effects on human health. Heliyon. 2023;**9**(9):e19496. DOI: 10.1016/j.heliyon.2023.e19496

[10] Singh A, Singh KK. An overview of the environmental and health consequences of air pollution. Iranica Journal of Energy & Environment. 2022;**13**(3):231-237

[11] Moronkeji A, Akinbo FO. Genotoxic response and histological alterations in rat lungs exposed to gasoline generator exhaust. Sultan Qaboos University Journal for Science [SQUJS]. 2024;**29**(1):15-27. Available from: https://journals.squ.edu.om/index.php/squjs/article/view/5780

[12] Duan RR, Hao K, Yang T. Air pollution and chronic obstructive pulmonary disease. Chronic Diseases and Translational Medicine. 2020;**6**(4):260-269

[13] Tahery N, Dobaradarn S, Zarea K, Cheraghi M, Mohammadi MJ, Hatamzadeh N, et al. Chronic obstructive pulmonary disease (COPD) and air pollution: A review. Jundishapur Journal of Chronic Disease Care. 2021;**10**(1):1-7

[14] de Santos UP, Arbex MA, Braga ALF, Mizutani RF, Cançado JED, Terra-Filho M, et al. Environmental air pollution: Respiratory effects. Jornal Brasileiro de Pneumologia. 2021;**47**(1):1-13

[15] Yang L, Zhang Y, Qi W, Zhao T, Zhang L, Zhou L, et al. Adverse effects of PM2. 5 on cardiovascular diseases. Reviews on Environmental Health. 2022;**37**(1):71-80. DOI: 10.1515/reveh-2020-0155

[16] Losacco C, Perillo A. Particulate matter air pollution and respiratory impact on humans and animals. Environmental Science and Pollution Research. 2018;**25**(34):33901-33910

[17] Miller MR, Newby DE. Air pollution and cardiovascular disease: Car sick. Cardiovascular Research. 2020;**116**(2):279-294

[18] Combes A, Franchineau G. Fine particle environmental pollution and cardiovascular diseases. Metabolism. 2019;**100**:153944

[19] Haberzettl P, O'Toole TE, Bhatnagar A, Conklin DJ. Exposure to fine particulate air pollution causes vascular insulin resistance by inducing pulmonary oxidative stress. Environmental Health Perspectives. 2016;**124**(12):1830-1839

[20] Bhatnagar A. Cardiovascular effects of particulate air pollution. Annual Review of Medicine. 2022;**73**:393-406

[21] Fu P, Guo X, Cheung FMH, Yung KKL. The association between PM 2.5 exposure and neurological disorders: A systematic review and meta-analysis. Science of the Total Environment. 2019;**655**:1240-1248. DOI: 10.1016/j.scitotenv.2018.11.218

[22] Block ML, Calderón-Garcidueñas L. Air pollution: Mechanisms of neuroinflammation and CNS disease. Trends in Neurosciences. 2009;**32**(9):506-516

[23] Calderón-Garcidueñas L, Torres-Jardón R, Kulesza RJ, Park SB, D'Angiulli A. Air pollution and detrimental effects on children's brain. The need for a multidisciplinary approach to the issue complexity and challenges. Frontiers in Human Neuroscience. 2014;**8**:1-7

[24] Haghani A, Morgan TE, Forman HJ, Finch CE. Air pollution neurotoxicity in the adult brain: Emerging concepts from experimental findings. Journal of Alzheimer's Disease. 2020;**76**(3):773-797

[25] Huang W, Xu H, Wu J, Ren M, Ke Y, Qiao J. Toward cleaner air and better health: Current state, challenges, and priorities. Science. 2024;**385**(6707):386-390

[26] Rentschler J, Leonova N. Global air pollution exposure and poverty. Nature Communications. 2023;**14**(1):1-11

[27] Folberth GA, Jones CD, O'Connor FM, Gedney N, Griffiths PT, Wiltshire AJ. Drivers of persistent changes in the global methane cycle under aggressive mitigation action. npj Climate and Atmospheric Science. 2025;**8**(1):136

[28] Feng JC, Yan J, Wang Y, Yang Z, Zhang S, Liang S, et al. Methane mitigation: Learning from the natural marine environment. Innovations. 2022;**3**(5):100297. DOI: 10.1016/j.xinn.2022.100297

[29] Runkov RA, Ilyasov DV. Spatial variability of methane emissions from soils of wet forests: A brief review. Environmental Dynamics and Global Climate Change. 2024;**14**(3):167-180

[30] Matsushita M, Ishikawa S, Magara K, Sato Y, Kimura H. The potential for ch4 production by syntrophic microbial communities in diverse deep aquifers associated with an accretionary prism and its overlying sedimentary layers. Microbes and Environments. 2020;**35**(1):ME19103

[31] Conrad R. Importance of hydrogenotrophic, aceticlastic and methylotrophic methanogenesis for methane production in terrestrial, aquatic and other anoxic environments: A mini review. Pedosphere. 2020;**30**(1):25-39. DOI: 10.1016/S1002-0160(18)60052-9

[32] Mar KA, Unger C, Walderdorff L, Butler T. Beyond CO_2 equivalence: The impacts of methane on climate, ecosystems, and health. Environmental Science & Policy. 2022;**134**(March):127-136. DOI: 10.1016/j.envsci.2022.03.027

[33] Thakur S, Solanki H. Role of methane in climate change and options for mitigation-a brief review. International Association of Biologicals and Computational Digest. 2022;**1**(2):275-281

[34] Cael BB, Goodwin PA. Global methane pledge versus carbon dioxide emission reduction. Environmental Research Letters. 2023;**18**(10):104015

[35] Sawyer WJ, Genina I, Brenneis RJ, Feng H, Li Y, Lennon Luo SLX. Methane emissions and global warming: Mitigation technologies, policy ambitions, and global efforts. MIT Science Policy Review. 2022;**3**:73-84

[36] Liu C, Chen D, Chen X. Atmospheric reactivity of methoxyphenols: A review. Environmental Science & Technology. 2022;**56**(5):2897-2916

[37] Su H, Cheng Y, Pöschl U. New multiphase chemical processes influencing atmospheric aerosols, air quality, and climate in the anthropocene. Accounts of Chemical Research. 2020;**53**(10):2034-2043

[38] Nguyen DH, Lin C, Vu CT, Cheruiyot NK, Nguyen MK, Le TH, et al. Tropospheric ozone and NO_x: A review of worldwide variation and meteorological influences. Environmental Technology and Innovation. 2022;**28**(x):102809

[39] McDuffie EE, Sarofim MC, Raich W, Jackson M, Roman H, Seltzer K, et al. The social cost of ozone-related mortality impacts from methane emissions. Earth's Future. 2023;**11**(9):e2023EF003853

[40] Fu Z, Xie HB, Elm J, Guo X, Fu Z, Fu Z, et al. Formation of low-volatile products and unexpected high formaldehyde yield from the atmospheric oxidation of methylsiloxanes. Environmental Science & Technology. 2020;**54**(12):7136-7145

[41] Nault BA, Jo DS, McDonald BC, Campuzano-Jost P, Day DA, Hu W, et al. Secondary organic aerosols from anthropogenic volatile organic compounds contribute substantially to air pollution mortality. Atmospheric Chemistry and Physics. 2021;**21**(14):11201-11224

[42] Stevenson DS, Zhao A, Naik V, O'Connor FM, Tilmes S, Zeng G, et al. Trends in global tropospheric hydroxyl radical and methane lifetime since 1850 from AerChemMIP. Atmospheric Chemistry and Physics. 2020;**20**(21):12905-12920

[43] Liu M, Song Y, Matsui H, Shang F, Kang L, Cai X, et al. Enhanced

atmospheric oxidation toward carbon neutrality reduces methane's climate forcing. Nature Communications. 2024;**15**(1):3-10

[44] Zhao Y, Saunois M, Bousquet P, Lin X, Berchet A, Hegglin MI, et al. Influences of hydroxyl radicals (OH) on top-down estimates of the global and regional methane budgets. Atmospheric Chemistry and Physics. 2020;**20**(15):9525-9546

[45] Alahmari A. Blood-brain barrier overview: Structural and functional correlation. Neural Plasticity. 2021;**2021**(1):6564585

[46] Benz F, Liebner S. Structure and function of the blood-brain barrier (BBB). In: Physiology, Pharmacology and Pathology of the Blood-Brain Barrier. Cham: Springer International Publishing; 2020. pp. 3-31

[47] Xiao M, Xiao ZJ, Yang B, Lan Z, Fang F. Blood-brain barrier: More contributor to disruption of central nervous system homeostasis than victim in neurological disorders. Frontiers in Neuroscience. 2020;**14**(August):1-17

[48] Greene C, Hanley N, Campbell M. Blood-brain barrier associated tight junction disruption is a hallmark feature of major psychiatric disorders. Translational Psychiatry. 2020;**10**(1):373. DOI: 10.1038/s41398-020-01054-3

[49] Paz AA, González-Candia A. Potential pharmacological target of tight junctions to improve the BBB permeability in neonatal hypoxic-ischemic encephalopathy diseases. Biochemical Pharmacology. 2023;**207**:115356. Available from: https://www.sciencedirect.com/science/article/abs/pii/S0006295222004518?via%3Dihub

[50] Ehsanifar M, Banihashemian B, Ehsanifar M. Exposure to air pollution nanoparticles: Oxidative stress and neuroinflammation. Journal of Biomedical Research & Environmental Sciences. 2021;**2**(10):964-976

[51] Alhussaini AR, Aljabri MR, Al-Harbi ZT, Almohammadi GA, Al-Harbi TM, Bashir S. Air pollution and its adverse effects on the central nervous system. Cureus. 2023;**15**(5):e38927. DOI: 10.7759/cureus.38927

[52] Bello-Medina PC, Rodríguez-Martínez E, Prado-Alcalá RA, Rivas-Arancibia S. Ozone pollution, oxidative stress, synaptic plasticity, and neurodegeneration. Neurología (English Edition). 2022;**37**(4):277-286

[53] Marin-Castañeda LA, Gonzalez-Garibay G, Garcia-Quintana I, Pacheco-Aispuro G, Rubio C. Mechanisms of ozone-induced neurotoxicity in the development and progression of dementia: A brief review. Frontiers in Aging Neuroscience. 2024;**16**(October):1-7

[54] Ramírez-Mendoza AA, Mendoza-Magaña ML, Ramírez-Herrera MA, Hernández-Nazara ZH, Domínguez-Rosales JA. Nitrooxidative stress and neuroinflammation caused by air pollutants are associated with the biological markers of neurodegenerative diseases. Antioxidants. 2024;**13**(3):326

[55] Juhász L, Tallósy SP, Nászai A, Varga G, Érces D, Boros M. Bioactivity of inhaled methane and interactions with other biological gases. Frontiers in Cell and Development Biology. 2022;**9**(January):1-8

[56] Chung KF, McGarvey L, Song WJ, Chang AB, Lai K, Canning BJ, et al. Cough hypersensitivity and chronic cough. Nature Reviews Disease

Primers. 2022;**8**(1):45. DOI: 10.1038/s41572-022-00370-w

[57] Lee YG, Lee PH, Choi SM, An MH, Jang AS. Effects of air pollutants on airway diseases. International Journal of Environmental Research and Public Health. 2021;**18**(18):9905. DOI: 10.3390/ijerph18189905

[58] He L, Weschler CJ, Morrison G, Li F, Zhang Y, Bergin MH, et al. Synergistic effects of ozone reaction products and fine particulate matter on respiratory pathophysiology in children with asthma. ACS ES&T Air. 2024;**1**(8):918-926

[59] Tu H, Hu Y, Hu K, Dong P, Wen Y, Jiang J, et al. Assessment of the respiratory disease mortality risk from single and composite exposures to PM2.5 and ozone — Guangzhou City, Guangdong Province, China, 2018-2021. China CDC Weekly. 2024;**6**(34):857-861

[60] Gangwar RS, Bevan GH, Palanivel R, Das L, Rajagopalan S. Oxidative stress pathways of air pollution mediated toxicity: Recent insights. Redox Biology. 2020;**34**:101545. DOI: 10.1016/j.redox.2020.101545

[61] Brinchmann BC, Skuland T, Rambøl MH, Szoke K, Brinchmann JE, Gutleb AC, et al. Lipophilic components of diesel exhaust particles induce pro-inflammatory responses in human endothelial cells through AhR-dependent pathway(s). Particle and Fibre Toxicology. 2018;**15**(1):1-17

[62] Janaszak-Jasiecka A, Siekierzycka A, Płoska A, Dobrucki IT, Kalinowski L. Endothelial dysfunction driven by hypoxia—The influence of oxygen deficiency on no bioavailability. Biomolecules. 2021;**11**(7):982. DOI: 10.3390/biom11070982

[63] Jimenez MTB, Hahad O, Kuntic M, Daiber A, Münzel T. Noise, air, and heavy metal pollution as risk factors for endothelial dysfunction. European Cardiology Review. 2023;**18**:e09

[64] Wang Z, Tang M. Research progress on toxicity, function, and mechanism of metal oxide nanoparticles on vascular endothelial cells. Journal of Applied Toxicology. 2021;**41**(5):683-700

[65] Kaur R, Singh V, Kumari P, Singh R, Chopra H, Emran TB. Novel insights on the role of VCAM-1 and ICAM-1: Potential biomarkers for cardiovascular diseases. Annals of Medicine and Surgery. 2022;**84**:December 2022. DOI: 10.1016/j.amsu.2022.104802

[66] Theofilis P, Sagris M, Oikonomou E, Antonopoulos AS, Siasos G, Tsioufis C, et al. Inflammatory mechanisms contributing to endothelial dysfunction. Biomedicine. 2021;**9**(7):1-21

[67] Calabriso N, Massaro M, Scoditti E, Carluccio C, Verri T, Carluccio MA. Epigenetic mechanisms in vascular inflammation: Modulation of endothelial adhesion molecules and endothelium-leukocyte adhesion. Frontiers in Bioscience-Landmark. 2023;**28**(9):194

[68] Lao XQ, Bo Y, Chen D, Zhang K, Szeto CC. Environmental pollution to kidney disease: An updated review of current knowledge and future directions. Kidney International. 2024;**106**(2):214-225. Available from: https://www.sciencedirect.com/science/article/abs/pii/S0085253824003442

[69] Okoye OC, Carnegie E, Mora L. Air pollution and chronic kidney disease risk in oil and gas- situated communities: A systematic review and meta-analysis. International Journal of Public Health. 2022;**67**:1604522

[70] Afsar B, Elsurer Afsar R, Kanbay A, Covic A, Ortiz A, Kanbay M. Air pollution and kidney disease: Review of current evidence. Clinical Kidney Journal. 2019;**12**(1):19-32

[71] Akinpelu M, Gamade SM, Akinbo F, Adeniyi TD, Elizebeth AF, Obeagu EI. Histopathological and biochemical effect of vitamin C and D on phosphine-induced hepatotoxicity in wistar rats. Asian Journal of Dental and Health Science. 2023;**3**(2):18-22

[72] Wang B, Li ZL, Zhang YL, Wen Y, Gao YM, Liu BC. Hypoxia and chronic kidney disease. eBioMedicine. 2022;**77**:103942. DOI: 10.1016/j.ebiom.2022.103942

[73] Piko N, Bevc S, Hojs R, Ekart R. The role of oxidative stress in kidney injury. Antioxidants. 2023;**12**(9):1772

[74] Tirichen H, Yaigoub H, Xu W, Wu C, Li R, Li Y. Mitochondrial reactive oxygen species and their contribution in chronic kidney disease progression through oxidative stress. Frontiers in Physiology. 2021;**12**(April):1-12

[75] Ho HJ, Shirakawa H. Oxidative stress and mitochondrial dysfunction in chronic kidney disease. Cells. 2023;**12**(1):1-29

[76] Aranda-Rivera AK, Cruz-Gregorio A, Aparicio-Trejo OE, Pedraza-Chaverri J. Mitochondrial redox signaling and oxidative stress in kidney diseases. Biomolecules. 2021;**11**(8):1-29

[77] Zhang Y, Tang C, Liu Y, Jiang H, Lu J, Lu Z, et al. Long-term ozone exposure is negatively associated with estimated glomerular filtration rate in Chinese middle-aged and elderly adults. Chemosphere. 2023;**341**:140040. Available from: https://www.sciencedirect.com/science/article/abs/pii/S004565352302310X

[78] Peng S, Chen B, Li Z, Sun J, Liu F, Yin X, et al. Ambient ozone pollution impairs glucose homeostasis and contributes to renal function decline: Population-based evidence. Ecotoxicology and Environmental Safety. 2024;**269**(December 2023):115803. DOI: 10.1016/j.ecoenv.2023.115803

[79] Lousa I, Reis F, Santos-Silva A, Belo L. The signaling pathway of TNF receptors: Linking animal models of renal disease to human CKD. International Journal of Molecular Sciences. 2022;**23**(6):3284

[80] Schroth J, Thiemermann C, Henson SM. Senescence and the aging immune system as major drivers of chronic kidney disease. Frontiers in Cell and Development Biology. 2020;**8**(October):1-10

[81] Xu X, Nie S, Ding H, Hou FF. Environmental pollution and kidney diseases. Nature Reviews Nephrology. 2018;**14**(5):313-324. DOI: 10.1038/nrneph.2018.11

[82] Berakdar N, Alahmad A. Review of oxidative stress and antioxidative. Innovare Journal of Life Sciences. 2022;**10**(1):1-3

[83] Aranda-Rivera AK, Cruz-Gregorio A, Arancibia-Hernández YL, Hernández-Cruz EY, Pedraza-Chaverri J. RONS and oxidative stress: An overview of basic concepts. Oxygen. 2022;**2**(4):437-478

[84] Moloney JN, Cotter TG. ROS signalling in the biology of cancer. Seminars in Cell & Developmental Biology. 2018;**80**:50-64. DOI: 10.1016/j.semcdb.2017.05.023

[85] Sies H, Jones DP. Reactive oxygen species (ROS) as pleiotropic physiological

signalling agents. Nature Reviews Molecular Cell Biology. 2020;**21**(7):363-383. DOI: 10.1038/s41580-020-0230-3

[86] Jomova K, Raptova R, Alomar SY, Alwasel SH, Nepovimova E, Kuca K, et al. Reactive oxygen species, toxicity, oxidative stress, and antioxidants: Chronic diseases and aging. In: Archives of Toxicology. Vol. 97. Berlin Heidelberg: Springer; 2023. pp. 2499-2574. DOI: 10.1007/s00204-023-03562-9

[87] Mailloux RJ. An update on methods and approaches for interrogating mitochondrial reactive oxygen species production. Redox Biology. 2021;**45**:102044

[88] Andrés Juan C, de la Lastra JMP, Plou FJ, Pérez-Lebeña E, Reinbothe S. Molecular sciences the chemistry of reactive oxygen species (ROS) revisited: Outlining their role in biological macromolecules (DNA, lipids and proteins) and induced pathologies. International Journal of Molecular Sciences. 2021;**22**:4642. DOI: 10.3390/ijms

[89] Moldogazieva NT, Zavadskiy SP, Astakhov DV, Terentiev AA. Lipid peroxidation: Reactive carbonyl species, protein/DNA adducts, and signaling switches in oxidative stress and cancer. Biochemical and Biophysical Research Communications. 2023;**687**:149167. Available from:https://www.sciencedirect.com/science/article/abs/pii/S0006291X23012615?via%3Dihub

[90] Adeniyi TD, Moronkeji A, Ralph-Okhiria OF, Wiqoyah N, Obi A, Maharani S, et al. Ascorbate and alpha-tocopherol mitigate toxic pathological changes in adult wistar rats exposed to cypermethrin. Folia Medica Indonesiana. 2023;**59**(4):387-395

[91] Adeniyi T, Moronkeji A, Fikayomi A. Histological and biochemical evaluation of the protective potential of ascorbate and alpha-tocopherol against cypermethrin-induced toxicity. The Journal of Experimental Life Science. 2024;**14**(1):33-39

[92] Jena AB, Samal RR, Bhol NK, Duttaroy AK. Cellular Red-Ox system in health and disease: The latest update. Biomedicine & Pharmacotherapy. 2023;**162**:114606. DOI: 10.1016/j.biopha.2023.114606

[93] Iancu Greceanu IM, Tănase HI, Dobrescu CM, Dorobăț ML. The mechanism and biochemistry in oxidative stress a brief overview. Current Trends in Natural Sciences. 2023;**12**(24):84-96

[94] Ursini F, Maiorino M. Lipid peroxidation and ferroptosis: The role of GSH and GPx4. Free Radical Biology & Medicine. 2020;**152**(December 2019):175-185

[95] Allowitz K, Taylor J, Harames K, Yoo J, Baloch O, Ramana KV. Oxidative stress-mediated lipid peroxidation-derived lipid aldehydes in the pathophysiology of neurodegenerative diseases. Current Neuropharmacology. 2024;**23**(6):23-25. Available from: http://www.ncbi.nlm.nih.gov/pubmed/39440770

[96] Li Y, Zhao T, Li J, Xia M, Li Y, Wang X, et al. Oxidative stress and 4-hydroxy-2-nonenal (4-HNE): Implications in the pathogenesis and treatment of aging-related diseases. Journal of Immunology Research. 2022;**2022**:2233906

[97] Sharma S, Sharma P, Bailey T, Bhattarai S, Subedi U, Miller C, et al. Electrophilic aldehyde 4-hydroxy-2-nonenal mediated signaling and mitochondrial dysfunction. Biomolecules. 2022;**12**(11):1555

[98] Žarković N, Gęgotek A, Łuczaj W, Jaganjac M, Šunjić SB, Žarković K, et al. Overview of the lipid peroxidation measurements in patients by the enzyme-linked immunosorbent assay specific for the 4-hydroxynonenal-protein adducts (4-HNE-ELISA). Frontiers in Bioscience-Landmark. 2024;**29**(4):153

[99] Kehm R, Baldensperger T, Raupbach J, Höhn A. Protein oxidation - Formation mechanisms, detection and relevance as biomarkers in human diseases. Redox Biology. 2021;**42**:101901

[100] Bandookwala M, Sengupta P. 3-nitrotyrosine: A versatile oxidative stress biomarker for major neurodegenerative diseases. The International Journal of Neuroscience. 2020;**130**(10):1047-1062. DOI: 10.1080/00207454.2020.1713776

[101] Demasi M, Augusto O, Bechara EJH, Bicev RN, Cerqueira FM, Da Cunha FM, et al. Oxidative modification of proteins: From damage to catalysis, signaling, and beyond. Antioxidants Redox Signal. 2021;**35**(12):1016-1080

[102] Paquim AMC. 8-oxoguanine and 8-oxodeoxyguanosine biomarkers of oxidative DNA damage: A review on HPLC–ECD determination. Molecules. 2022;**27**(5):1-18

[103] Jelic MD, Mandic AD, Maricic SM, Srdjenovic BU. Oxidative stress and its role in cancer. Journal of Cancer Research and Therapeutics. 2021;**17**(7):22-28

[104] Graille M, Wild P, Sauvain JJ, Hemmendinger M, Canu IG, Hopf NB. Urinary 8-OHDG as a biomarker for oxidative stress: A systematic literature review and meta-analysis. International Journal of Molecular Sciences. 2020;**21**(11):1-24

[105] Rochette L, Dogon G, Rigal E, Zeller M, Cottin Y, Vergely C. Lipid peroxidation and iron metabolism: Two corner stones in the homeostasis control of ferroptosis. International Journal of Molecular Sciences. 2023;**24**(1):1-17

[106] Iuchi K, Takai T, Hisatomi H. Cell death via lipid peroxidation and protein aggregation diseases. Biology (Basel). 2021;**10**(5):399

[107] Andrés CMC, Pérez de la Lastra JM, Andrés Juan C, Plou FJ, Pérez-Lebeña E. Impact of reactive species on amino acids—Biological relevance in proteins and induced pathologies. International Journal of Molecular Sciences. 2022;**23**(22):14049

[108] Petushkova AI, Zamyatnin AA. Redox-mediated post-translational modifications of proteolytic enzymes and their role in protease functioning. Biomolecules. 2020;**10**(4):650

[109] Poetsch AR. The genomics of oxidative DNA damage, repair, and resulting mutagenesis. Computational and Structural Biotechnology Journal. 2020;**18**:207-219. DOI: 10.1016/j.csbj.2019.12.013

[110] Kumar N, Gupta H. Chapter 18 - Methane: Risk assessment, environmental, and health hazard. In: Singh J, Kaushik RD, Chawla M, editors. Hazardous Gases. Academic Press; 2021. pp. 225-238. ISBN 9780323898577. Available from: www.sciencedirect.com/science/article/pii/B9780323898577000098

[111] Ainslie A, Huiting W, Barazzuol L, Bergink S. Genome instability and loss of protein homeostasis: Converging paths to neurodegeneration? Open Biology. 2021;**11**(4):200296

[112] Tortajada-Pérez J, del Carranza AV, Trujillo-del Río C, Collado-Pérez M,

Millán JM, García-García G, et al. Lipid oxidation at the crossroads: Oxidative stress and neurodegeneration explored in caenorhabditis elegans. Antioxidants. 2025;**14**(1):1-27

[113] Pisoschi AM, Pop A. The role of antioxidants in the chemistry of oxidative stress: A review. European Journal of Medicinal Chemistry. 2015;**97**:55-74. DOI: 10.1016/j.ejmech.2015.04.040

[114] Jomova K, Alomar SY, Alwasel SH, Nepovimova E, Kuca K, Valko M. Several lines of antioxidant defense against oxidative stress: Antioxidant enzymes, nanomaterials with multiple enzyme-mimicking activities, and low-molecular-weight antioxidants. In: Archives of Toxicology. Vol. 98. Berlin Heidelberg: Springer; 2024. pp. 1323-1367. DOI: 10.1007/s00204-024-03696-4

[115] Gusti AMT, Qusti SY, Alshammari EM, Toraih EA, Fawzy MS. Antioxidants-related superoxide dismutase (SOD), catalase (CAT), glutathione peroxidase (GPX), glutathione-s-transferase (GST), and nitric oxide synthase (NOS) gene variants analysis in an obese population: A preliminary case-control study. Antioxidants. 2021;**10**(4):595

[116] Cecerska-Heryć E, Surowska O, Heryć R, Serwin N, Napiontek-Balińska S, Dołęgowska B. Are antioxidant enzymes essential markers in the diagnosis and monitoring of cancer patients – A review. Clinical Biochemistry. 2021;**93**(November 2020):1-8

[117] Lievanos-Ruiz FJ, Fenton-Navarro B. Enzymatic biomarkers of oxidative stress in patients with depressive disorders. A systematic review. Clinical Biochemistry. 2024;**130**:110788. Available from: https://www.sciencedirect.com/science/article/abs/pii/S0009912024000821

[118] Eddaikra A, Eddaikra N. Endogenous enzymatic antioxidant defense and pathologies. In: Antioxidants - Benefits, Sources, Mechanisms of Action. London, UK, London: IntechOpen; 2021. Available from: https://www.intechopen.com/books/advanced-biometric-technologies/liveness-detection-in-biometrics

[119] de Alves AF, de Moura AC, Andreolla HF, da Veiga ABG, Fiegenbaum M, Giovenardi M, et al. Gene expression evaluation of antioxidant enzymes in patients with hepatocellular carcinoma: RT-qPCR and bioinformatic analyses. Genetics and Molecular Biology. 2021;**44**:e20190373

[120] Allameh A, Niayesh-Mehr R, Aliarab A, Sebastiani G, Pantopoulos K. Oxidative stress in liver pathophysiology and disease. Antioxidants. 2023;**12**(9):1-23

[121] Moezzi D, Dong Y, Jain RW, Lozinski BM, Ghorbani S, D'Mello C, et al. Expression of antioxidant enzymes in lesions of multiple sclerosis and its models. Scientific Reports. 2022;**12**(1):1-13. DOI: 10.1038/s41598-022-16840-w

[122] Roy Z, Bansal R, Siddiqui L, Chaudhary N. Understanding the role of free radicals and antioxidant enzymes in human diseases. Current Pharmaceutical Biotechnology. 2023;**24**(10):1265-1276

[123] Sidorova Y, Domanskyi A. Detecting oxidative stress biomarkers in neurodegenerative disease models and patients. Methods and Protocols. 2020;**3**(4):1-14

[124] Mello LD. Potential contribution of ELISA and LFI assays to assessment of the oxidative stress condition based on 8-oxodG biomarker. Analytical Biochemistry. 2021;**628**(January):114215. DOI: 10.1016/j.ab.2021.114215

[125] Li J, Pan L, Pan W, Li N, Tang B. Recent progress of oxidative stress associated biomarker detection. Chemical Communications. 2023;**59**:7361-7374

[126] Zhou X, Gao S, Yue M, Zhu S, Liu Q, Zhao XE. Recent advances in analytical methods of oxidative stress biomarkers induced by environmental pollutant exposure. TrAC Trends in Analytical Chemistry. 2023;**160**:116978. Available from: https://www.sciencedirect.com/science/article/abs/pii/S0165993623000651

[127] Zhang A, Luo X, Li Y, Yan L, Lai X, Yang Q, et al. Epigenetic changes driven by environmental pollutants in lung carcinogenesis: A comprehensive review. Frontiers in Public Health. 2024;**12**:1420933

[128] Olmedo-Suárez MÁ, Ramírez-Díaz I, Pérez-González A, Molina-Herrera A, Coral-García MÁ, Lobato S, et al. Epigenetic regulation in exposome-induced tumorigenesis: Emerging roles of ncRNAs. Biomolecules. 2022;**12**(4):1-47

[129] Santibáñez-Andrade M, Quezada-Maldonado EM, Rivera-Pineda A, Chirino YI, García-Cuellar CM, Sánchez-Pérez Y. The road to malignant cell transformation after particulate matter exposure: From oxidative stress to genotoxicity. International Journal of Molecular Sciences. 2023;**24**(2):1782

[130] Quezada-Maldonado EM, Sánchez-Pérez Y, Chirino YI, García-Cuellar CM. Airborne particulate matter induces oxidative damage, DNA adduct formation and alterations in DNA repair pathways. Environmental Pollution. 2021;**287**:117313. Available from: https://www.sciencedirect.com/science/article/abs/pii/S0269749121008952

[131] Reed L, Arlt VM, Phillips DH. The role of cytochrome P450 enzymes in carcinogen activation and detoxication: An in vivo-in vitro paradox. Carcinogenesis. 2018;**39**(7):851-859

[132] Bukowska B, Duchnowicz P, Tumer TB, Michałowicz J, Krokosz A. Precarcinogens in food – Mechanism of action, formation of DNA adducts and preventive measures. Food Control. 2023;**152**:109884. Available from: https://www.sciencedirect.com/science/article/abs/pii/S0956713523002840

[133] Fiorilla I, Martinotti S, Todesco AM, Bonsignore G, Cavaletto M, Patrone M, et al. Chronic inflammation, oxidative stress and metabolic plasticity: Three players driving the pro-tumorigenic microenvironment in malignant mesothelioma. Cells. 2023;**12**(16):2048

[134] Yu W, Tu Y, Long Z, Liu J, Kong D, Peng J, et al. Reactive oxygen species bridge the gap between chronic inflammation and tumor development. Oxidative Medicine and Cellular Longevity. 2022;**2022**:2606928

[135] Zhou W, Yang L, Nie L, Lin H. Unraveling the molecular mechanisms between inflammation and tumor angiogenesis. American Journal of Cancer Research. 2021;**11**(2):301-317

[136] García-Guede Á, Vera O, Ibáñez-de-Caceres I. When oxidative stress meets epigenetics: Implications in cancer development. Antioxidants. 2020;**9**(6):1-26

[137] Ilango S, Paital B, Jayachandran P, Padma PR, Nirmaladevi R. Genetic and epigenetic alterations in cancer. Frontiers in Bioscience. 2020;**25**(6):1058-1109

[138] Arias-Pérez RD, Taborda NA, Gómez DM, Narvaez JF, Porras J, Hernandez JC. Inflammatory effects of particulate matter air pollution.

Environmental Science and Pollution Research. 2020;**27**(34):42390-42404

[139] Ofremu GO, Raimi BY, Yusuf SO, Dziwornu BA, Nnabuife SG, Eze AM, et al. Exploring the relationship between climate change, air pollutants and human health: Impacts, adaptation, and mitigation strategies. Green Energy and Resources. 2025;**3**(2):100074. DOI: 10.1016/j.gerr.2024.100074. Available from: https://www.sciencedirect.com/science/article/pii/S2949720524000286

[140] Krug JD, Riedel TP, Lewandowski M, Lonneman WA, Turlington JM, Zavala J, et al. Mutagenic atmospheres generated from the photooxidation of NO_x with selected VOCs and a complex mixture: Apportionment of aromatic mutagenicity for reacted gasoline vapor. Atmospheric Environment. 2024;**334**:120668. Available from: https://www.sciencedirect.com/science/article/abs/pii/S1352231024003431

[141] Hrdina AIH, Kohale IN, Kaushal S, Kelly J, Selin NE, Engelward BP, et al. The parallel transformations of polycyclic aromatic hydrocarbons in the body and in the atmosphere. Environmental Health Perspectives. 2022;**130**(2):1-9

[142] Vondráček J, Machala M. The role of metabolism in toxicity of polycyclic aromatic hydrocarbons and their non-genotoxic modes of action. Current Drug Metabolism. 2021;**22**(8):584-595

[143] Rosendahl Huber A, Van Hoeck A, Van Boxtel R. The mutagenic impact of environmental exposures in human cells and cancer: Imprints through time. Frontiers in Genetics. 2021;**12**(October):1-14

[144] Kumari S, Sharma S, Advani D, Khosla A, Kumar P, Ambasta RK. Unboxing the molecular modalities of mutagens in cancer. Environmental Science and Pollution Research. 2022;**29**(41):62111-62159

[145] Chatterjee N, Walker GC. Mechanisms of DNA damage, repair, and mutagenesis. Environmental and Molecular Mutagenesis. 2017;**58**(5):235-263. DOI: 10.1002/em.22087

[146] Ernst L, Steinfeld B, Barayeu U, Klintzsch T, Kurth M, Grimm D, et al. Methane formation driven by reactive oxygen species across all living organisms. Nature. 2022;**603**:482-487

[147] Keppler F, Boros M, Polag D. Radical-driven methane formation in humans evidenced by exogenous isotope-labeled DMSO and methionine. Antioxidants. 2023;**12**(7):1-22

[148] Comba A, Faisal SM, Varela ML, Hollon T, Al-Holou WN, Umemura Y, et al. Uncovering spatiotemporal heterogeneity of high-grade gliomas: From disease biology to therapeutic implications. Frontiers in Oncology. 2021;**11**(August):1-16

[149] Fischer EG. Nuclear morphology and the biology of cancer cells. Acta Cytologica. 2020;**64**(6):511-519

[150] Torrisi F, Alberghina C, D'aprile S, Pavone AM, Longhitano L, Giallongo S, et al. The hallmarks of glioblastoma: Heterogeneity, intercellular crosstalk and molecular signature of invasiveness and progression. Biomedicine. 2022;**10**(4):806

[151] Gilles HC, Garbutt T, Landrum J. Hepatocellular carcinoma. Critical Care Nursing Clinics of North America. 2022;**34**(3):289-301. Available from: https://www.sciencedirect.com/science/article/abs/pii/S0899588522000314

[152] Vij M, Calderaro J. Pathologic and molecular features of hepatocellular

carcinoma: An update. World Journal of Hepatology. 2021;**13**(4):393-410

[153] Kefeli A, Basyigit S, Yeniova AO. Diagnosis of hepatocellular carcinoma. In: Updates in Liver Cancer. London, UK: IntechOpen; 2017

[154] Kew MC. Hepatocellular carcinoma in developing countries: Prevention, diagnosis and treatment. World Journal of Hepatology. 2012;**4**(3):99-104. DOI: 10.4254/wjh.v4.i3.99

[155] Che M, Yin R. Analysis of the global burden of ovarian cancer in adolescents. International Journal of Gynecological Cancer. 2025;**35**(4):101620. DOI: 10.1016/j.ijgc.2024.101620

[156] Bajaj A. Vortex and crater-squamous cell carcinoma lung. Journal of Medical Case Studies. 2023;**1**(1):1-3

[157] Khamar B. Desmocollin-3 and cancer. Biomedical Journal of Scientific Technical Research. 2017;**1**(5):1511-1515

[158] Savci-Heijink CD, Kosari F, Aubry MC, Caron BL, Sun Z, Yang P, et al. The role of desmoglein-3 in the diagnosis of squamous cell carcinoma of the lung. The American Journal of Pathology. 2009;**174**(5):1629-1637

[159] Galindo I, Gómez-Morales M, Díaz-Cano I, Andrades Á, Caba-Molina M, Miranda-León MT, et al. The value of desmosomal plaque-related markers to distinguish squamous cell carcinoma and adenocarcinoma of the lung. Upsala Journal of Medical Sciences. 2020;**125**(1):19-29. DOI: 10.1080/03009734.2019.1692101

[160] Dong Y, Li S, Sun X, Wang Y, Lu T, Wo Y, et al. Desmoglein 3 and keratin 14 for distinguishing between lung adenocarcinoma and lung squamous cell carcinoma. OncoTargets and Therapy. 2020;**13**:11111-11124

[161] Kumari N, Bhandari S, Ishfaq A, Butt SRR, Ekhator C, Karski A, et al. Primary cardiac angiosarcoma: A review. Cureus. 2023;**15**(7):e41947. DOI: 10.7759/cureus.41947

[162] Zhao B, Zhu H, Buja LM, Garcia RA. Primary cardiac angiosarcoma with giant cells: A case report of an autopsy. Human Pathology: Case Reports. 2020;**20**(February):200370. DOI: 10.1016/j.ehpc.2020.200370

[163] Lalonde K, Dumoulin E, Povitz M. Case of pericardial angiosarcoma in a 42-year-old male presenting with cardiac tamponade, pleural effusions, and dyspnea on the pulmonary service. Chest. 2021;**160**(4):A2382. DOI: 10.1016/j.chest.2021.07.2060

[164] Sozzi FB, Gnan E, Pandolfi A, Iacuzio L, Kim JK, Canetta C, et al. Diagnostic algorithm using multimodal imaging for the differential diagnosis of intra-cardiac masses. Journal of Clinical Medicine. 2025;**14**(2):508

[165] Mubasher M, Hanafi A, Syed T, Zambon M, Hoefen R. The great masquerader: Cardiac angiosarcoma mimicking thrombotic thrombocytopenia purpura. Journal of the American College of Cardiology. 2020;**75**(11):2655. DOI: 10.1016/S0735-1097(20)33282-4

[166] Satoshi H, Kyoko Y, Shuji I, Atsushi I, Noriko H, Shinichiro S, et al. A case of pericardial angiosarcoma with refractory pericardial tamponade treated with multidisciplinary therapy with pericardial fenestration, radiotherapy and chemotherapy. International Journal of Case Reports and Images (IJCRI). 2015;**6**(11):707

[167] Nezami BG, MacLennan GT. Clear cell renal cell carcinoma: A comprehensive review of its histopathology, genetics, and differential diagnosis. International Journal of Surgical Pathology. 2025;**33**(2):265-280

[168] Zvirblyte J, Nainys J, Juzenas S, Goda K, Kubiliute R, Dasevicius D, et al. Single-cell transcriptional profiling of clear cell renal cell carcinoma reveals a tumor-associated endothelial tip cell phenotype. Communications Biology. 2024;**7**(1):780. DOI: 10.1038/s42003-024-06478-x

Chapter 4

Methanol-To-Hydrogen Conversion with Integrated Heat Recovery: A Pathway to Sustainable Energy Systems

Kun Cheng, Yunlin Shao and Weidong Chen

Abstract

This chapter explores the innovative integration of methanol reforming and heat recovery technologies for sustainable hydrogen production, building on the techno-economic and multi-objective optimization framework presented in the author's prior research. The focus is on converting methanol-to-hydrogen efficiently while maximizing energy utilization through cascading heat recovery systems. The chapter highlights the role of proton-exchange membrane fuel cells in enhancing system performance and reducing carbon emissions. By leveraging advanced optimization techniques, the proposed system demonstrates significant improvements in energy efficiency and cost-effectiveness, offering a viable pathway for decarbonizing energy systems. The discussion also addresses the challenges and opportunities in scaling up methanol-to-hydrogen technologies, emphasizing their potential to contribute to a sustainable energy future. This work aligns with global efforts to innovate methanol utilization and transition toward cleaner energy solutions.

Keywords: methanol reforming, hydrogen production, heat recovery, proton-exchange membrane fuel cell, techno-economic assessment, multi-objective optimization, sustainable energy systems, cascading energy utilization, carbon emissions reduction, energy efficiency

1. Introduction

1.1 Background and rational of methanol-to-hydrogen conversion

The continued reliance on fossil fuels intensifies global energy shortages while worsening environmental problems like climate change, caused by increasing greenhouse gas emissions [1]. In response, nations worldwide are prioritizing renewable energy solutions, building low-carbon energy systems, and improving energy utilization efficiency. Hydrogen has become a key energy vector in this transition, valued for its exceptional energy density (120–142 MJ/kg—almost triple that of gasoline) and its clean combustion properties, producing only water when used in fuel cells or burned [2, 3].

Among hydrogen-based technologies, fuel cell combined heat and power (FC-CHP) systems have gained prominence for their exceptional efficiency (reaching 80–90% in combined generation mode) and suitability for distributed energy applications. FC-CHP systems generate both electricity and usable heat from a single fuel source, typically natural gas, using a fuel cell [4]. Leading economies, including China, the U.S., and the EU, have endorsed FC-CHP systems as a strategic solution for achieving carbon neutrality and strengthening energy security [5]. This shift underscores the necessity of replacing fossil fuels with hydrogen, a clean, storable, and adaptable energy carrier [6].

A major obstacle to hydrogen adoption, however, is its storage and transportation, complicated by its low volumetric energy density, high flammability, and the expensive processes required for compression or liquefaction. To address these challenges, decentralized hydrogen production has emerged as a practical solution. Current small-scale hydrogen generation methods include solar-driven processes [7–9], electrolysis (electrocatalytic water splitting) [10, 11], and fossil-derived feedstocks like methanol and methane [12]. Notably, methanol-to-hydrogen conversion—especially through steam methanol reforming (SMR)—is the most advanced and widely adopted method, responsible for nearly 95% of global hydrogen production [13]. Enhancing this process with improved energy recovery, lower emissions, and cost-effective operations is essential for advancing sustainable energy systems.

1.2 Literature review of methanol-to-hydrogen conversion systems

The fuel reforming unit (FRU) serves as a critical component in localized hydrogen generation, transforming hydrocarbon (C_nH_{2n+2}) and alcohol-based ($C_nH_{2n+2}O_n$) feedstocks—whether renewable or fossil-derived—into hydrogen, reducing the hazards linked to its large-scale transport [14]. To achieve high-purity hydrogen, purification techniques such as pressure swing adsorption (PSA) and palladium-based membrane (Pd-M) separation are widely utilized [15, 16]. For instance, Rodriges et al. [17] designed a combined methanol/dimethyl ether reforming system paired with a proton-exchange membrane fuel cell (PEMFC), incorporating a Pd-M for hydrogen refinement. Similarly, Wang et al. [18] proposed an SMR-PEMFC setup where methanol combustion provided the necessary heat for reforming, while a Pd-M maintained hydrogen purity. In another study, Zhang et al. [19] incorporated a PSA-based CO_2 removal system into a PEMFC to minimize CO_2-induced performance degradation. Nevertheless, the overall electrical efficiency of FRU-FC configurations remains constrained (40–60%) due to substantial heat dissipation during hydrogen synthesis, purification, and electricity generation.

1.2.1 Integration of waste heat recovery for enhanced system efficiency

To enhance system efficiency, recent studies have investigated waste heat utilization and cogeneration strategies, demonstrating significant performance gains. For instance, Perna et al. [20] developed an ammonia-based CHP configuration that attained 81% overall energy efficiency by effectively capturing thermal output, compensating for modest electrical conversion rates. Similarly, Chang et al. [21] demonstrated a thermally integrated PAFC-methanol reformer setup with enhanced vaporization, achieving 277 mW/cm^2 power density through sophisticated thermal management. Another innovative approach by Li et al. [22] combined a gas turbine with MSR and PEMFC technologies, where exhaust gases simultaneously powered the reformer and provided domestic hot water, resulting in 81.8% combined and 50.5% electrical efficiency.

Alternative thermal sources have also been explored to optimize hydrogen production. Chen's team [23] incorporated geothermal energy into a micro-CCHP system for methanol reforming, achieving 66.3% system efficiency. Meanwhile, Wang et al. [24] developed a solar-enhanced MSR-PAFC configuration with thermal absorption capabilities, delivering seasonal efficiencies of 73.7% (cooling) and 51.7% (heating). Addressing diverse operational needs, Wu et al. [25] integrated hybrid heat pumps with PEMFC-CHP systems, improving electrical output by 7.9–20.3% depending on thermal mode. While conventional reformers still predominantly use fossil fuel combustion—producing moisture-laden (10–20% water content) exhaust streams [26]—the substantial latent energy in these emissions remains largely unexploited. Current advancements highlight the efficacy of heat-integrated methanol-to-hydrogen systems while underscoring the necessity for cleaner thermal inputs and innovative heat recovery methods to meet environmental objectives.

1.2.2 Integration of waste heat recovery for power generation

Cutting-edge research has turned to advanced thermal recovery methods to enhance the power output of methanol-based hydrogen generation systems. Among the most promising solutions is the incorporation of Organic Rankine Cycle (ORC) units, which effectively transform waste heat into supplemental electrical energy. For example, Li's research' team [27] created an integrated MSR-PEMFC-ORC configuration that harnesses thermal energy from heat transfer oil, reaching 38.52% exergetic efficiency after system refinement. Parallel work by Chang and colleagues [28] examined a tri-generation setup merging PEMFC with ORC and refrigeration components, revealing that strategic fluid choice (such as R601) could push exergy efficiency to 46–47% across seasonal variations. These findings emphasize how precise thermal management can extract maximum value from process waste streams.

The scientific community has investigated complementary heat recovery approaches to push efficiency boundaries further. Sulaiman et al. [29] conducted empirical tests on TEG-enhanced PEMFC arrangements, discovering that recovering even 1% of dissipated heat demands 45 thermoelectric modules, highlighting the difficulties in low-grade thermal energy conversion. Conversely, Lu's group [30] reported a 4.3% electrical efficiency boost by combining PEMFC with ORC technology, while Lee's innovative design [31] achieved a 10% system-wide improvement through coordinated heat capture from both the fuel cell and reformer. The inherent temperature constraints of PEMFC operation (typically 60–80°C) [32] present fundamental limitations for ORC performance, driving research toward novel thermal conversion methods or multi-technology integrations.

This body of work demonstrates that strategic heat recovery implementation, particularly through ORC solutions, can substantially increase electrical yields from hydrogen production systems. However, the effective utilization of low-temperature waste streams continues to pose technical hurdles. Emerging research directions emphasize the development of combined recovery systems and the adoption of high-temperature fuel cell variants to optimize energy extraction from thermal byproducts.

1.2.3 Economic analysis and optimization for improved system configurations

Economic considerations have become paramount in advancing methanol-based hydrogen production systems toward commercial feasibility. Contemporary studies reveal how intelligent system design and optimization can dramatically improve cost-effectiveness while boosting operational performance. A breakthrough study by Sarabchi's team [33] achieved dual benefits by combining a Kalina cycle with high-temperature PEMFC technology, utilizing waste heat to drive the thermodynamic cycle. Their EES software-assisted optimization yielded impressive results—a 17.72% decrease in production costs coupled with a 29.3% gain in exergetic performance versus conventional setups. Parallel research by Zheng et al. [34] implemented comprehensive process integration in methanol-reforming PEMFC systems, delivering 4.5% lower electricity costs and 2.22% better exergy efficiency than standard CCHP configurations.

Modern optimization techniques have shown particular promise in economic enhancement. Xi and collaborators [35] applied multi-objective genetic algorithms to refine a 5 kW residential PEMFC-CCHP system, achieving an optimized annual operating cost of $29,337 with 39.9% exergy efficiency. Zhao's research group [36] demonstrated additional gains by incorporating thermal recovery through adsorption chillers and dehumidifiers, ultimately reaching $143,000 annual cost at 33.97% exergy efficiency after algorithmic optimization. Further advancements came from Li's work [27], combining ORC with HT-PEMFC, where optimization reduced energy costs by 6.47%, while Yu's innovative Organic Flash Cycle (OFC) integration [37] identified R245fa and isobutene as optimal fluids with hourly costs below $31.

Current system configurations still face energy utilization challenges. ORC subsystems typically capture just 10–20% of PEMFC waste heat, while substantial thermal energy remains untapped in reformer exhaust streams. These limitations highlight critical opportunities for improving primary energy conversion efficiency through more sophisticated heat recovery approaches.

2. Case study presentation for methanol-to-hydrogen conversion with integrated heat recovery systems

2.1 System descriptions

Figure 1(a) presents the integrated MSR-PEMFC-ORC-LTES tri-generation system, which synergistically combines methanol steam reforming with organic Rankine cycle and latent thermal energy storage technologies. The system architecture transforms waste heat from both the reformer and fuel cell into valuable electricity, thermal energy, and distilled water. At the core of hydrogen production lies a network of heat exchangers, catalytic reactors, combustion chambers, and pressure swing adsorption units. The process initiates with preheating and vaporization of methanol solution through thermal exchange with hot reformate gases. Subsequent superheating occurs via combustion exhaust, elevating the vapor temperature to the optimal 280°C reforming condition. The resulting hydrogen-rich stream undergoes progressive cooling before entering the purification stage, where pressure swing adsorption isolates high-purity hydrogen for PEMFC operation. Electrochemical conversion in the fuel cell generates power while producing exhaust streams that pass through the thermal storage unit before emission. The energy recovery subsystem features an organic Rankine cycle with standard components, including vaporization chambers,

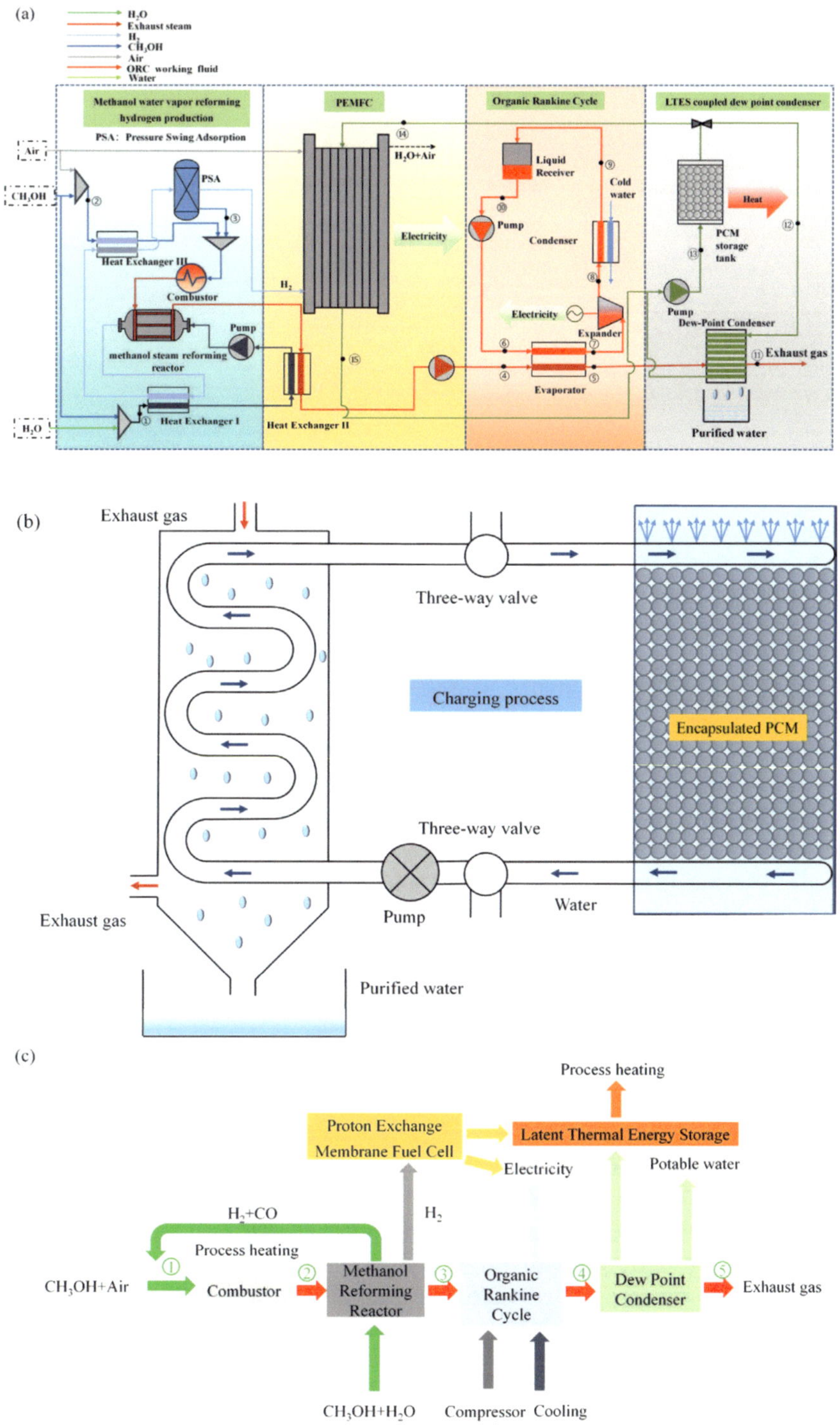

Figure 1.
(a) Process flow diagram of the MSR-PEMFC-ORC-LTES tri-generation system; (b) structure of LTES coupled dew point condenser; (c) energy conversion schematic of the tri-cogeneration system.

Working fluids	Environmental data		Physical data	
	ODP	GWP	Critical temperature (K)	Critical pressure (kPa)
R245fa	0	1030	427.16	3651.00
R600	0	~20	425.13	3796.00
R152a	0	124	386.41	4516.75
R113	1	6130	487.23	3436
R500	0.738	8100	378.65	4455
R12	1	10,890	385.15	4114

Table 1.
Properties of the working fluids.

expansion turbines, and condensation equipment. Thermal energy from exhaust streams drives the working fluid through its phase changes, producing additional electricity *via* the turbine-generator set. Fluid selection critically influences both thermodynamic performance and environmental compliance, with current options including R600, R152a, and R245fa—chosen for their favorable environmental profiles and established commercial availability as referenced in studies [38, 39], as shown in **Table 1**.

The thermal recovery system incorporates a phase-change energy storage unit integrated with a condensation heat exchanger, as illustrated in **Figure 1(b)**. This configuration captures residual thermal energy from the ORC subsystem, transferring it to a composite phase-change material blend of expanded graphite and paraffin wax for storage. During colder periods, the accumulated thermal energy is utilized for water heating applications. The system employs water as the thermal transfer medium due to its favorable thermal properties, safety characteristics, and economic viability. The condensation process continues until the exhaust gases reach saturation temperature, yielding substantial quantities of distilled water through condensation. **Figure 1(c)** presents the comprehensive energy flow diagram for the integrated tri-generation system.

2.2 Mathematical model and experiment

This section presents integrated mathematical frameworks that simultaneously address thermodynamic performance, economic viability, and environmental factors for both the ORC subsystem and the phase-change thermal storage integrated condensation unit. The developed analytical models enable systematic evaluation of capital expenditures, ecological footprint, and exergetic performance of the thermal recovery system. Reaction kinetics modeling for the reforming process is implemented through computational tools, with governing equations solved using MATLAB coupled with REFPROP NIST thermodynamic libraries, while process simulation leverages the Aspen Plus platform. Critical operational parameters for the thermal recovery components are specified in **Table 2**. A genetic algorithm-based multi-criteria optimization approach, previously validated in our research [43], is applied to determine optimal system configurations.

Mathematical formulation incorporates several simplifying considerations to enhance computational efficiency. The analysis assumes continuous steady-state operation throughout the tri-generation system [44]. Fluid transport phenomena

DOI: http://dx.doi.org/10.5772/intechopen.1011625

Symbol	Definition	Unit	Value
m_{ex}	Mass flow of exhaust gases	kg/s	
T_4	Initial temperature of exhaust gases	K	
P_{ex}	Pressure of exhaust gases	kPa	
d_4	Initial moisture of exhaust gases	$g\ /\ kg(dryair)$	
T_{cin}	Cold water inlet temperature of the condenser	K	298.15
T_{12}	PCM tank outlet temperature	K	325.15
T_{13}	PCM tank inlet temperature	K	338.15
T_0	Air temperature	K	301.15 [40]
P_0	Air pressure	kPa	101 [40]
n	Life time	years	20 [41]
c_p	Specific heat capacity of water at constant pressure	$kJ\,kg^{-1}K^{-1}$	4.2 [42]
k	Heat transfer coefficient of the heat exchanger	$W\ m^{-2}\ K^{-1}$	800

Table 2.
Operating and design conditions of the heat recovery system.

neglect both frictional pressure losses and thermal dissipation in distribution networks [45]. The energy balance equations exclude contributions from macroscopic motion and positional energy terms [45]. For the thermal storage subsystem, the thermodynamic analysis disregards temperature variations in the heat transfer fluid during pumping operations [46].

2.2.1 Model development methanol steam reforming reaction

The methanol steam reforming process involves three interrelated chemical transformations occurring simultaneously: steam reforming (SR), thermal decomposition (MD), and water-gas shift (WGS) reactions. These transformations are represented by the following stoichiometric relationships [26]:

$$CH_3OH + H_2O \rightarrow 3H_2 + CO_2\left(\Delta H_{298_K} = 49.4\ kJ\ /\ mol\right) \tag{1}$$

$$CH_3OH \rightarrow 2H_2 + CO\left(\Delta H_{298_K} = 90.5\ kJ\ /\ mol\right) \tag{2}$$

$$CO + H_2O \leftrightarrow H_2 + CO_2\left(\Delta H_{298_K} = -41.1\ kJ\ /\ mol\right) \tag{3}$$

The product stream contains not only hydrogen but also residual CO, CO_2, water vapor, and unreacted methanol. The CuO/ZrO_2 catalytic system enhances reaction kinetics while suppressing undesirable side reactions [47]. Reaction conditions, particularly temperature and the molar ratio of water-to-methanol, critically influence conversion efficiency. Studies indicate that elevated water-to-methanol ratios enable satisfactory conversion even at reduced temperatures [14]. However, insufficient water content may lead to catalyst deactivation through coking.

Process simulation employs a Gibbs reactor approach within UniSim Design software [48], where system thermodynamics are governed by minimization of total Gibbs free energy. This is mathematically described by:

$$G_{total} = \Sigma\left(n_i G_i^o\right) + RT_{msr}\,\Sigma\left(N_i \ln\left(f_i / f_i^o\right)\right) \tag{4}$$

Here, G_{total} represents the system's total Gibbs free energy, G_i^0 denotes standard-state Gibbs free energy for component i, while n_i and f_i correspond to molar quantity and fugacity, respectively. The term f_i^0 indicates standard-state fugacity, and N represents the total number of system components.

2.2.2 Thermodynamic model of ORC

The thermodynamic analysis of the ORC subsystem models the evaporator and condenser as countercurrent heat exchange devices. Thermal energy from exhaust streams is transferred to the working fluid in the evaporator, producing superheated vapor that drives the expansion turbine to generate electrical power. The system's thermal performance is quantified through fundamental energy balance equations [46], where the heat recovery capacity (q_{orc}) relates to exhaust enthalpy change, while the working fluid mass flow (m_{orc}) depends on the evaporator's energy input. Electrical output (W_e) derives from the turbine's isentropic expansion of the working fluid, with an assumed turbine efficiency of 85%.

Pumping requirements account for the working fluid's pressure elevation, expressed as a function of mass flow, pressure differential, fluid density, and pump efficiency (83% assumed). Net power production considers both turbine output and pumping work. The condenser's thermal load (q_{con}) is determined by the working fluid's enthalpy change during phase transition, dictating the required cooling water flow rate.

Heat exchanger sizing follows conventional thermal design methodology [49], where required surface areas (A_{eva}, A_{con}) are calculated using logarithmic mean temperature differences (ΔT_m) and overall heat transfer coefficients. The temperature correction factor (F) incorporates heat capacity ratios and heat transfer effectiveness through the number of transfer units (NTU) parameters, ensuring accurate thermal performance prediction under varying operating conditions.

2.2.3 LTES coupled dew point condenser stage

The thermal storage system employs a cylindrical PCM containment structure with height-to-diameter unity to minimize environmental heat dissipation [43]. The integrated dew point condenser captures residual thermal energy from ORC exhaust streams, with total recoverable heat being a function of enthalpy changes in both reformer and fuel cell exhaust flows. Water circulation requirements for heat extraction are determined by the system's thermal load and temperature differentials across the storage unit.

Pumping energy consumption depends on water flow characteristics and pressure requirements, incorporating fluid density and pump efficiency parameters. The condenser's heat transfer surface area is sized according to thermal performance criteria involving overall heat transfer coefficients and logarithmic mean temperature differences [46]. The PCM tank's energy storage capacity accounts for charging duration while deducting environmental thermal losses, which are calculated using tank surface area, thermal resistance, and temperature gradients. Geometric relationships determine the tank's surface area based on its volume, while the required storage volume itself is derived from the PCM's thermophysical properties, including phase-change characteristics, density, specific heat capacities, and latent heat of fusion. The condensation process yields distilled water as exhaust gases cool to saturation conditions, with production rates determined by moisture content differentials and inert gas flow rates. Water collection quantities are calculated based on humidity ratios at condenser inlet and outlet conditions, considering the mass flow of non-condensable gases in the exhaust stream.

2.2.4 Exergy analysis

Thermodynamic limitations dictate that heat-to-work conversion processes inevitably incur energy degradation. Practical systems experience exergy destruction through various irreversibilities, making exergy analysis essential for determining maximum achievable work output. The specific exergy at any state ($E_{x\,i}$) is expressed in terms of enthalpy and entropy deviations from ambient reference conditions (h_o, s_o), incorporating both thermal and entropic components. For open systems operating under steady-state conditions, the exergy balance relates input and output exergy flows to work interactions and system losses. This fundamental relationship accounts for all exergy transfers across system boundaries, including both thermal and mechanical energy contributions. System performance is ultimately characterized by the exergetic efficiency metric, which quantifies the fraction of available energy that avoids destruction during the conversion process.

The efficiency expression normalizes the avoidable losses against the net exergy input to the system, providing a meaningful measure of thermodynamic perfection. This formulation properly accounts for all work interactions while isolating the irreversible components of energy conversion.

2.2.5 Economic model

A comprehensive techno-economic assessment requires evaluating both thermodynamic performance and financial feasibility. Beyond exergy analysis, this study incorporates capital and operational expenditure models to determine system viability. The annual operating expenses account for electricity consumption during off-peak charging periods, calculated as a function of power input and regional electricity tariffs. Power requirements aggregate all pumping demands within the thermal recovery subsystems.

Capital investments are annualized using standard financial engineering methods, incorporating interest rates and equipment lifespans through capital recovery factors. Maintenance costs are proportionally estimated relative to initial investments. The total capital outlay comprises the sum of individual component costs, with detailed cost functions for each major subsystem provided in **Table 3**. This dual-focus approach enables simultaneous optimization of both thermodynamic and economic performance metrics [44, 46]. The complete economic evaluation combines these annualized capital and operational expenditures into a unified metric representing the system's total yearly cost. This integrated approach enables direct comparison

System component	Cost function	Reference
Expander	$C_{\exp} = 3.380 \times (q_{orc} - q_{con}) \times \left(1 + \left(\frac{0.05}{1 - \eta_{\exp}}\right)^3\right)$	[49]
Flat plate heat exchanger	$C_{hx} = 268 + 517 \times A_{hx}$	[50]
Pump	$C_{pump} = 705 \times W_{pump}^{0.71}\left(1 + \frac{0.2}{1 - \eta_{pump}}\right)$	[50]
PCM tank	$C_{tank} = 8.67 \times \times 10^{\left[2.92 \times \exp(0.143 \times \log V_{LTES})\right]} + \varepsilon \times V_{LTES} \times \rho_{pcm} \times 1.19$	[50]
Dew point condenser	$C_{dew} = 1379 \times A_{dew}^{0.8}$	[51]

Table 3.
The cost function for the entire components.

between different system configurations and facilitates identification of the cost-optimal design.

2.2.6 Environmental model

The environmental benefits of the integrated energy system are quantified through a comparative analysis of greenhouse gas emissions. The assessment focuses on carbon dioxide reduction achieved by displacing conventional fossil fuel-based electricity and heat production. Emission savings are determined by applying standardized conversion factors to the net energy output, accounting for both electrical generation and recovered thermal energy after compensating for system parasitic losses. The analysis employs a baseline emission coefficient of 0.7 kg CO_2 equivalent per kilowatt-hour derived from regional grid averages [46].

System performance is evaluated through comprehensive energy utilization metrics that consider multiple output streams. The overall efficiency calculation incorporates net electrical production from both power generation subsystems, useful thermal energy stored in the phase-change materials, and the recovered latent heat from water condensation. These outputs are normalized against the total chemical energy input from methanol combustion, represented by its lower heating value. The thermal contribution of distilled water production is included through enthalpy difference calculations relative to ambient conditions [22].

2.3 Results and discussions

This analysis presents the comparative multi-objective optimization results for the heat recovery system using three candidate working fluids under uniform operating conditions. The optimization framework balances three key performance indicators—total annualized cost, exergetic efficiency, and carbon reduction potential—as visualized in the three-dimensional Pareto frontiers shown in **Figure 2**. The optimization targets simultaneously maximize system efficiency (η) and emission reductions (mGHG) while minimizing expenditures (Ctotal), with the theoretical ideal solution and practical optimum clearly marked.

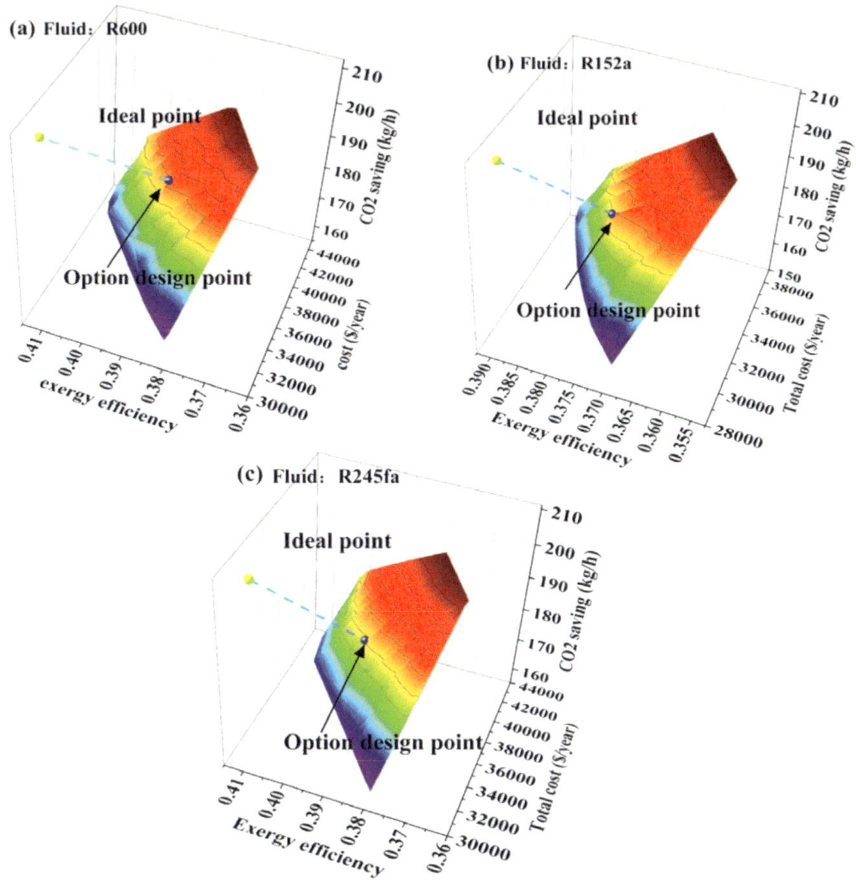

Figure 2.
Pareto frontier of the heat recovery system with (a) R600, (b) R152a, and (c) R245fa.

Table 4 reveals comparable exergetic performance across all working fluids (R600: 0.388, R245fa: 0.39, R152a: 0.41) due to effective heat recovery in the condensation subsystem. However, the ORC subsystem demonstrates notable fluid-dependent variations, with R600 achieving 67% higher exergy efficiency than R152a. Economic analysis shows modest cost differences (9% variation) between options, while emission reductions remain within 3.6% of each other. Power generation capacity exhibits more significant divergence, with R152a producing 33% less electricity than R600 and R245fa, making the latter two clearly preferable for electrical output.

The integrated system achieves 81% overall energy efficiency—a 22% improvement over conventional alternatives. Component-level cost breakdowns in **Figure 3(a)** and **(c)** reveal that the thermal storage subsystem requires two to three times greater capital investment than the ORC unit. Remarkably, the condensation process contributes 95% of the system's total emission reductions by effectively capturing exhaust gas latent heat, though this comes with increased cost implications. Distilled water production reaches 216.03 kg/h, further enhancing the system's resource recovery capabilities (**Table 4**).

		R600	R152a	R245fa
Design parameters	T_5 (K)	377.35	376.92	374.55
	ΔT_e (K)	1	1	1
	ΔT_c (K)	8.69	7.24	7.61
	T_{cout} (K)	303.15	303.15	303.15
	T_{11} (K)	334.96	336.15	335.14
Performance variables	η (–)	0.388	0.41	0.39
	m_{GHG} (kg/h)	191.61	184.85	188.84
	C_{total} ($/year)	36,108	32,794	36,196
	W_e (kW)	15.34	10.54	15.64

Table 4.
Results of multi-objective optimization.

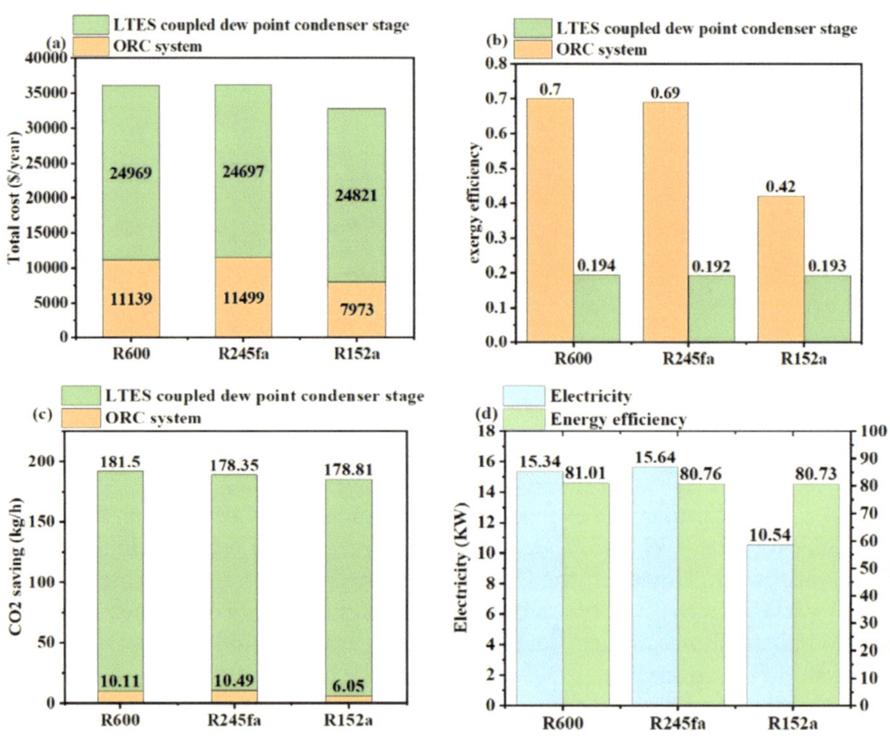

Figure 3.
(a) Cost distribution, (b) exergy efficiency of each stage, (c) CO_2 emission reduction of each stage, (d) electricity of ORC system and energy efficiency of tri-generation system.

2.4 Key conclusion points of the case study analysis

This study presents an innovative hybrid energy system that synergistically combines proton-exchange membrane fuel cell technology with methanol steam reforming, augmented by organic Rankine cycle and latent thermal energy storage components. The integrated configuration enables simultaneous production of electrical power, thermal energy, and distilled water while addressing key challenges in renewable energy integration. A novel composite phase-change material, engineered through the combination of expanded graphite and industrial paraffin, demonstrates exceptional thermal conductivity properties for efficient energy storage applications. The system's performance is rigorously optimized across economic, environmental, and thermodynamic dimensions using advanced multi-objective genetic algorithms, with detailed parametric analysis conducted on both the hydrogen production and energy recovery subsystems. Key findings from this comprehensive investigation include:

The methanol reforming process achieves optimal performance at specific operating parameters, delivering sufficient hydrogen output to meet the fuel cell's power generation requirements. Comparative analysis of working fluids reveals marginal differences in cost and emissions performance but significant variations in electrical output capacity, with certain refrigerants demonstrating superior thermodynamic characteristics. The thermal storage subsystem contributes disproportionately to the system's environmental benefits while exhibiting different efficiency characteristics compared to the power generation components.

The developed composite thermal storage material exhibits remarkable thermophysical properties that significantly enhance system performance. When implemented in the complete tri-generation configuration, the system demonstrates exceptional energy conversion efficiency and output capacity, substantially outperforming conventional geothermal-assisted systems in terms of overall energy utilization. These results validate the technical and economic viability of the proposed integrated energy solution for residential-scale applications.

3. Conclusions

This section examines recent advancements in methanol-based hydrogen generation systems incorporating thermal energy recovery technologies, highlighting a novel proton-exchange membrane fuel cell configuration with multi-stage energy utilization. The proposed system architecture shows marked improvements across multiple performance metrics including energy conversion efficiency, environmental impact reduction, economic viability, and sustainable energy integration.

3.1 Challenges

Significant technical obstacles must be addressed to realize the full potential of these integrated energy systems. Maintaining optimal thermal conditions for the heat-intensive reforming process while ensuring stable operation under variable loads presents considerable engineering difficulties. The intricate thermal balancing required for effective cascading energy recovery becomes particularly challenging when adapting to real-world fluctuating demand patterns, where even minor thermal inefficiencies can substantially diminish overall energy recovery capabilities.

Catalyst development remains a pivotal area requiring innovation, as current materials face persistent issues with thermal degradation, carbon accumulation, and chemical contamination. These challenges are compounded by the need to maintain stringent hydrogen purity standards for fuel cell operation, necessitating complex and energy-intensive gas purification processes. Parallel economic challenges stem from the substantial capital requirements associated with high-performance system components, underscoring the need for innovative financing approaches and standardized modular designs to facilitate commercial adoption.

3.2 Future perspectives

Future progress in this field appears promising through several key development pathways. The integration of renewable methanol synthesis with efficient reforming technologies could enable a carbon-neutral hydrogen infrastructure. Emerging computational techniques and artificial intelligence applications offer transformative potential for optimizing system performance across multiple scales, from molecular-level catalyst design to full-scale thermal management. Concurrent advances in materials science may yield breakthrough components with enhanced durability and efficiency. Realizing these technological possibilities will require complementary policy support and international cooperation to establish unified standards and accelerate technology deployment. The successful transition from research prototypes to commercial systems will depend on coordinated efforts spanning technical innovation, economic modeling, and regulatory framework development across multiple sectors.

Acknowledgements

The authors gratefully acknowledge the generous funding under the National Key Research and Development Program of China, 2023YFB4006002, to support the research activities for this presented work.

Nomenclature

A	area (m^2)
C	cost ($ /year)
CRF	CAPITAL recovery factor
C_P	specific heat ($kJ/(kg\,K)$)
CPCM	composite phase-change material
CHP	combined heat and power
d	moisture content ($g/kg(dry\ air)$)
E	energy efficiency
F	correction factor
FC-CHP	hydrogen fuel cell cogeneration system
FRU	fuel reforming unit
GT	gas turbine
h	enthalpy (kJ)
HTES	HIGH-temperature thermal energy storagek heat transfer coefficient(W/($m^2 \cdot$ K))
KC	Kalina cycle
LTES	latent-thermal energy storage
m	mass flow rate (kg/s)
M	fuzzy judgment matrix

DOI: http://dx.doi.org/10.5772/intechopen.1011625

MSR	methanol steam reforming
MSR-PAFC	phosphoric acid fuel cell
NTU	heat transfer unit
n	life time (years)
OF	objective function
OFC	Organic Flash Cycle
ORC	Organic Rankine Cycle
P	pressure (kPa)
PSA	pressure swing absorption
PEMFC	proton-exchange membrane fuel cell
Q	stored energy (kWh)
q	time rate of heat transfer
s	entropy (J/ (kg K))
T	temperature (K)
TEG	thermoelectric generator
V	volume
W	power (kJ)
w	weight
EG	expanded graphite
Greek letters	
ρ	density (kg / m^3)
ε	PCM volume fraction
η	efficiency
ã	latent heat (kJ/kg)
δ	maintenance factor
Δ	difference
Subscripts	
cin	condenser cold side inlet
cout	condenser cold side outlet
con	condenser
cap	capital cost
cm	capital and maintenance costs
dew	dew point condenser
e	electricity (kJ)
ex	EXHAUST steam
eva	evaporator
exp.	expander
f	fluid
hx	heat exchanger
in	inlet
max	maximum
min	minimum
net	net power (kJ)
nor	normalization
op	operational cost
orcin	inflow to ORC
orcout	outflow ORC
pump	pump
s	solid

Author details

Kun Cheng[1], Yunlin Shao[1*] and Weidong Chen[2*]

1 Key Laboratory of Low-grade Energy Utilization Technologies and Systems, Ministry of Education, Chongqing University, Chongqing, China

2 Department of Mechanical Engineering, National University of Singapore, Singapore

*Address all correspondence to: shaoyunlin@cqu.edu.cn; chenweidong@nus.edu.sg

References

[1] De Cian E, Sue Wing I. Global energy consumption in a warming climate. Environmental and Resource Economics. 2017;**72**(2):365-410

[2] Lotrič A, Sekavčnik M, Hočevar S. Effectiveness of heat-integrated methanol steam reformer and polymer electrolyte membrane fuel cell stack systems for portable applications. Journal of Power Sources. 2014;**270**:166-182

[3] Mah AXY, Ho WS, Hassim MH, Hashim H, Liew PY, Muis ZA. Targeting and scheduling of standalone renewable energy system with liquid organic hydrogen carrier as energy storage. Energy. 2021;**218**:119475

[4] Al Moussawi H, Fardoun F, Louahlia-Gualous H. Review of tri-generation technologies: Design evaluation, optimization, decision-making, and selection approach. Energy Conversion and Management. 2016;**120**:157-196

[5] Wang X, Shu G, Tian H, Wang R, Cai J. Operation performance comparison of CCHP systems with cascade waste heat recovery systems by simulation and operation optimisation. Energy. 2020;**206**:118123

[6] Buchalik R, Nowak G, Nowak I. Mathematical model of a thermoelectric system based on steady- and rapid-state measurements. Applied Energy. 2021;**293**:116943

[7] Razmi AR, Hanifi AR, Shahbakhti M. Design, thermodynamic, and economic analyses of a green hydrogen storage concept based on solid oxide electrolyzer/fuel cells and heliostat solar field. Renewable Energy. 2023;**215**:118996

[8] Razmi AR, Hanifi AR, Shahbakhti M. Techno-economic analysis of a novel concept for the combination of methane pyrolysis in molten salt with heliostat solar field. Energy. 2024;**301**:1311644

[9] Ahmad YH, Ibrahim MF, Banu JR, Al-Qaradawi SY. Recent advances on the use of promoters in biochemical hydrogen production: A comprehensive review. Energy Conversion and Management. 2024;**317**:118814

[10] Yao Y, Lyu J, Li X, Chen C, Verpoort F, Wang J, et al. A review of efficient electrocatalysts for the oxygen evolution reaction at large current density. DeCarbon. 2024;**5**:100062

[11] Khalil M, Dincer I. An investigation of hydrogen production using 3D-printed cathodes for alkaline water electrolysis. Energy Conversion and Management. 2024;**302**:118169

[12] Yang P, Zhang Y, Yang C, Chen J, Liu Z, Deng C, et al. Thermodynamic performance comparison of a SOFC system integrated with steam reforming and dry reforming by utilizing different fuels. Energy Conversion and Management. 2024;**300**:117981

[13] Li N, Cui X, Zhu J, Zhou M, Liso V, Cinti G, et al. A review of reformed methanol-high temperature proton exchange membrane fuel cell systems. Renewable and Sustainable Energy Reviews. 2023;**182**:113395

[14] Mei D, Qiu X, Liu H, Wu Q, Yu S, Xu L, et al. Progress on methanol reforming technologies for highly efficient hydrogen production and applications. International Journal of Hydrogen Energy. 2022;**47**(84):35757-35777

[15] Nordio M, Wassie SA, Annaland MVS, Tanaka DAP, Sole JLV, Gallucci F. Techno-economic evaluation on a hybrid technology for low hydrogen concentration separation and purification from natural gas grid. International Journal of Hydrogen Energy. 2021;**46**(45):23417-23435

[16] Bernardo G, Araújo T, da Silva Lopes T, Sousa J, Mendes A. Recent advances in membrane technologies for hydrogen purification. International Journal of Hydrogen Energy. 2020;**45**(12):7313-7338

[17] Rodrigues CT, de França Lopes G, Alonso CG, de Matos Jorge LM, Paraíso PR. An autonomous fuel cell: Methanol and dimethyl ether steam reforming direct fed to fuel cell. International Journal of Hydrogen Energy. 2023;**48**(10):4052-4063

[18] Wang Z, Mao J, He Z, Liang F. Energy-exergy analysis of an integrated small-scale LT-PEMFC based on steam methane reforming process. Energy Conversion and Management. 2021;**246**:114685

[19] Liu J, Jiang Y, Zhang X, Fu L, Deng M. Performance optimization of an HT-PEMFC and PSA integrated system with impure hydrogen containing CO_2. Applied Thermal Engineering. 2022;**214**:118859

[20] Perna A, Minutillo M, Jannelli E, Cigolotti V, Nam SW, Han J. Design and performance assessment of a combined heat, hydrogen and power (CHHP) system based on ammonia-fueled SOFC. Applied Energy. 2018;**231**:1216-1229

[21] Chang C-P, Wu Y-C, Chen W-Y, Pan C, Su Y-C, Huang Y-J, et al. A hybrid phosphorus-acid fuel cell system incorporated with oxidative steam reforming of methanol (OSRM) reformer. Renewable Energy. 2020;**153**:530-538

[22] Li C, Wang Z, Liu H, Guo F, Xiu X, Qin J, et al. 4E analysis of a novel proton exchange membrane fuel cell/engine based cogeneration system with methanol fuel for ship application. Energy. 2023;**282**:128741

[23] Chen X, Zhou H, Yu Z, Li W, Tang J, Xu C, et al. Thermodynamic and economic assessment of a PEMFC-based micro-CCHP system integrated with geothermal-assisted methanol reforming. International Journal of Hydrogen Energy. 2020;**45**(1):958-971

[24] Wang J, Wu J, Xu Z, Li M. Thermodynamic performance analysis of a fuel cell trigeneration system integrated with solar-assisted methanol reforming. Energy Conversion and Management. 2017;**150**:81-89

[25] Wu W, Zhai C, Sui Y, Zhang H. A novel distributed energy system using high-temperature proton exchange membrane fuel cell integrated with hybrid-energy heat pump. Energy Conversion and Management. 2021;**235**:113990

[26] Wang J, Zhang Y, Peng S, Yu S, Ran J, Chua KJ, et al. Thermal analysis of a micro tubular reactor for methanol steam reforming by optimizing the multilayer arrangement of catalyst bed for the catalytic combustion of methanolr. International Journal of Hydrogen Energy. 2023;**48**(73):28315-28332

[27] Li Y, Li D, Ma Z, Zheng M, Lu Z, Song H, et al. Performance analysis and optimization of a novel vehicular power system based on HT-PEMFC integrated methanol steam reforming and ORC. Energy. 2022;**257**:124729

[28] Chang H, Wan Z, Zheng Y, Chen X, Shu S, Tu Z, et al. Energy- and

exergy-based working fluid selection and performance analysis of a high-temperature PEMFC-based micro combined cooling heating and power system. Applied Energy. 2017;**204**:446-458

[29] Sulaiman MS, Singh B, Mohamed WANW. Experimental and theoretical study of thermoelectric generator waste heat recovery model for an ultra-low temperature PEM fuel cell powered vehicle. Energy. 2019;**179**:628-646

[30] Lu X, Du B, Zhu W, Yang Y, Xie C, Tu Z, et al. Thermodynamic and dynamic analysis of a hybrid PEMFC-ORC combined heat and power (CHP) system. Energy Conversion and Management. 2023;**292**:134865

[31] Lee JY, Lee JH, Kim TS. Thermo-economic analysis of using an organic Rankine cycle for heat recovery from both the cell stack and reformer in a PEMFC for power generation. International Journal of Hydrogen Energy. 2019;**44**(7):3876-3890

[32] Chitsaz A, Haghghi MA, Hosseinpour J. Thermodynamic and exergoeconomic analyses of a proton exchange membrane fuel cell (PEMFC) system and the feasibility evaluation of integrating with a proton exchange membrane electrolyzer (PEME). Energy Conversion and Management. 2019;**186**:487-499

[33] Sarabchi N, Mahmoudi SMS, Yari M, Farzi A. Exergoeconomic analysis and optimization of a novel hybrid cogeneration system: High-temperature proton exchange membrane fuel cell/Kalina cycle, driven by solar energy. Energy Conversion and Management. 2019;**190**:14-33

[34] Liang Z, Liang Y, Luo X, Wang H, Wu W, Chen J, et al. Integration and optimization of methanol-reforming proton exchange membrane fuel cell system for distributed generation with combined cooling, heating and power. Journal of Cleaner Production. 2023;**411**:137342

[35] Chen X, Zhou H, Li W, Yu Z, Gong G, Yan Y, et al. Multi-criteria assessment and optimization study on 5 kW PEMFC based residential CCHP system. Energy Conversion and Management. 2018;**160**:384-395

[36] Zhao J, Cai S, Huang X, Luo X, Tu Z. 4E analysis and multiobjective optimization of a PEMFC-based CCHP system with dehumidification. Energy Conversion and Management. 2021;**248**:114789

[37] Yu Z, Feng C, Lai Y, Xu G, Wang D. Performance assessment and optimization of two novel cogeneration systems integrating proton exchange membrane fuel cell with organic flash cycle for low temperature geothermal heat recovery. Energy. 2022;**243**:122725

[38] Wang S, Liu C, Zhang S, Li Q, Huo E. Multi-objective optimization and fluid selection of organic Rankine cycle (ORC) system based on economic-environmental-sustainable analysis. Energy Conversion and Management. 2022;**254**:115238

[39] Ding Y, Liu C, Zhang C, Xu X, Li Q, Mao L. Exergoenvironmental model of Organic Rankine Cycle system including the manufacture and leakage of working fluid. Energy. 2018;**145**:52-64

[40] Nami H, Mahmoudi SMS, Nemati A. Exergy, economic and environmental impact assessment and optimization of a novel cogeneration system including a gas turbine, a supercritical CO_2 and an organic Rankine cycle (GT-HRSG/

SCO2). Applied Thermal Engineering. 2017;**110**:1315-1330

[41] Nasser M, Megahed TF, Ookawara S, Hassan H. Performance evaluation of PV panels/wind turbines hybrid system for green hydrogen generation and storage: Energy, exergy, economic, and enviroeconomic. Energy Conversion and Management. 2022;**267**:115870

[42] Zhou F, Song Y, Ma G. Load characteristics analysis of the water source heat pump heating system using data center waste heat. Energy Conversion and Management. 2023;**296**:117679

[43] Shao YL, Soh KY, Islam MR, Chua KJ. Thermal, exergy and economic analysis of a cascaded packed-bed tank with multiple phase change materials for district cooling system. Energy. 2023;**268**:126746

[44] Wang S, Liu C, Tang J, Xiao T, Huo E, Guan Z. Multi-mode and exergoeconomic analysis of a novel combined cooling, heating, and power system applied in the geothermal field. Energy Conversion and Management. 2023;**276**:116565

[45] Zare V. A comparative exergoeconomic analysis of different ORC configurations for binary geothermal power plants. Energy Conversion and Management. 2015;**105**:127-138

[46] Shao YL, Soh KY, Wan YD, Huang ZF, Islam MR, Chua KJ. Multi-objective optimization of a cryogenic cold energy recovery system for LNG regasification. Energy Conversion and Management. 2021;**244**:114524

[47] Liao M, Qin H, Guo W, Gao P, Xiao H. Porous reticular CuO/ZnO/CeO_2/ZrO_2 catalyst derived from polyacrylic acid hydrogel system on Al_2O_3 foam ceramic support for methanol steam reforming microreactor. Ceramics International. 2021;**47**(23):33667-33677

[48] Wu Z, Xu G, Zhang W, Xue X, Chen H. Thermodynamic and economic analysis of a new methanol steam reforming system integrated with CO_2 heat pump and cryogenic separation system. Energy. 2023;**283**:128501

[49] Towler G, Sinnott R. Chemical Engineering Design: Principles, Practice and Economics of Plant and Process Design. Oxford: Butterworth-Heinemann; 2021

[50] Sanaye S, Shirazi A. Thermo-economic optimization of an ice thermal energy storage system for air-conditioning applications. Energy and Buildings. 2013;**60**:100-109

[51] Peng X, She X, Nie B, Li C, Li Y, Ding Y. Liquid air energy storage with LNG cold recovery for air liquefaction improvement. Energy Procedia. 2019;**158**:4759-4764

Chapter 5

Methanogens in Soils and Climate Mitigation: Community Dynamics, Functional Markers, and Applications

Gamze Gürerk and Sevcan Aydin

Abstract

The global climate crisis is reshaping ecosystem functioning, and methane exerts a disproportionately high warming influence. Methanogenic archaea, the sole biological producers of methane, operate under anoxic conditions and mediate carbon conversion processes that regulate both energy flow and the global carbon cycle. Much of the existing literature has focused on methane accounting or on the mitigating roles of methanotrophic bacteria, whereas comprehensive syntheses that jointly address methanogen diversity, adaptive responses to climate stressors, and biotechnological applications remain limited. The conceptual novelty of this work lies in redefining methanogens not merely as methane producers but as climate-mitigating ecosystem engineers that regulate carbon fluxes and climate feedback. Within this framework, the chapter outlines the biochemical pathways of methanogenesis, highlights the diagnostic value of the *mcrA* gene, and synthesizes patterns of community diversity, distribution, and succession across soil ecosystems. It further examines physiological and community-level adaptations to warming, moisture variability, and CO_2 enrichment, and evaluates opportunities and risks related to renewable biomethane production, wetland and peatland restoration, and greenhouse gas management. Framed within an interdisciplinary and sustainability-oriented perspective, the chapter links microbial ecology with environmental policy, offering a timely and comprehensive integrative synthesis rarely achieved in a single work. It also emphasizes the need to incorporate molecular-level microbial processes into broader climate and sustainability discussions. Ultimately, this chapter highlights the strategic roles of methanogens in clean energy, ecosystem resilience, and climate mitigation, and proposes research, monitoring, and policy priorities to ensure their ecological and technological potential is harnessed responsibly.

Keywords: methanogenic archaea, soil microbiome, climate change impacts, sustainable agriculture, microbial adaptation, methanogenesis

1. Introduction

The twenty-first century is witnessing one of the most pressing ecological challenges, defined largely by a global climate crisis triggered by human activity. This crisis is not limited to shifts in average temperatures; it is reshaping the very functioning of ecosystems. Methane (CH_4), which holds a central position in the global carbon cycle, is particularly noteworthy. Despite its lower atmospheric concentration compared to carbon dioxide, methane has nearly 84 times the global warming potential over a 20-year period [1–3]. According to NOAA data, the global average atmospheric methane concentration reached 1922.6 ppb in 2023, marking the highest level since measurements began. More recent observations indicate that by April 2025, methane concentrations had risen further to 1934.9 ppb [4, 5]. Climate modeling suggests that even a mere 1% increase in methane could drive a measurable rise in global surface temperatures [6]. These findings indicate that understanding the methane cycle is not simply a matter of academic curiosity, but a strategic necessity for addressing climate change.

At the biological core of this potent greenhouse gas lie methanogenic archaea, among the planet's oldest and most resilient microorganisms. From rice paddies and thawing permafrost layers to wetlands and lake sediments, these strict anaerobes thrive in oxygen-free habitats, carrying out methanogenesis, the only known biological pathway for methane production [7, 8]. Over their 3.5 billion year evolutionary history, they have consistently shaped Earth's carbon balance, acting as key agents in the long-term trajectory of the climate system.

The ecological role of methanogens, however, extends far beyond the methane production alone. At the final stage of anaerobic decomposition, they establish symbiotic relationships with fermentative bacteria and acetogens, converting hydrogen, acetate, and methyl compounds into methane. This metabolic cooperation ensures the thermodynamic viability of fermentation and stabilizes energy flow across ecosystems. In this sense, methanogens are not merely passive producers of methane but can also be seen as microbial ecosystem engineers that regulate the energy economy of their habitats [9].

This delicate balance is becoming increasingly fragile, as climate change reshapes the very environmental factors (temperature, moisture, pH, and substrate availability) that determine the structure of methanogenic communities. For instance, the rapid colonization of newly exposed soils by pioneer methanogens in thawing permafrost, the tendency of rising agricultural temperatures to favor hydrogenotrophic pathways, or the swift recovery of methanogenic activity within days after drought stress all indicate that these microorganisms are not merely the passive indicators of climate change. Instead, they respond dynamically to shifting conditions and, in some cases, may even influence those conditions themselves [10, 11].

This chapter explores methanogenic archaea across a broad spectrum from their biochemical mechanisms to their diversity in soil ecosystems, their strategies for adapting to climate stressors, and their potential in biotechnological applications. Unlike previous syntheses that primarily focused on methane fluxes or on the mitigating roles of methanotrophic bacteria, this chapter distinguishes itself by emphasizing soil-based methanogenic communities and using the *mcrA* gene as a unifying framework that links molecular mechanisms to ecosystem-scale processes. By integrating biochemical, ecological, and biotechnological perspectives, it offers a holistic and interdisciplinary understanding of methanogens, bridging microbial ecology with sustainability and climate mitigation research.

In doing so, this work represents one of the few integrative syntheses that connect molecular-level processes to ecosystem functioning and biotechnological applications.

The aim here is to fill this gap by evaluating methanogens not only in terms of their ecological functions but also as strategic players in climate solutions, approached from an interdisciplinary perspective. The conceptual novelty of this chapter lies in redefining methanogens not merely as methane producers but as climate-mitigating ecosystem engineers that regulate carbon fluxes and climate feedback. They are described as "engineers" because their metabolic activities and ecological interactions actively shape biogeochemical cycles and help maintain the functional stability of ecosystems. The literature synthesis in this chapter follows a narrative review approach combining insights from microbiology, soil science, and biotechnology. Sources were gathered from Scopus, Web of Science, and Google Scholar, focusing on peer-reviewed works published between 2006 and 2025. The search used keywords such as *"methanogenic archaea," "mcrA gene," "soil ecosystems,"* and *"biomethane production."* Emphasis was placed on studies addressing soil-based methanogenic communities, environmental stress responses, and biotechnological applications. This approach provides a transparent and integrative framework consistent with the sustainability-oriented scope of the chapter. Ultimately, the chapter seeks to reposition these ancient microorganisms, not simply as methane producers, but as ecological actors capable of shaping the course of the climate crisis and serving at the core of sustainable biotechnological solutions.

2. Biochemical, genetic, and ecological foundations of methanogens

Methanogens occur in a wide range of environments, from deep-sea sediments to permafrost layers, and they are the only microorganisms that convert carbon compounds into methane under anaerobic conditions. This metabolic capacity sustains their energy metabolism and defines their function as important ecosystem engineers that shape the global carbon cycle [7, 8]. Methane is produced through three main biochemical pathways: hydrogenotrophic, acetoclastic, and methylotrophic methanogenesis. The hydrogenotrophic route involves the reduction of CO_2 by H_2 and commonly prevails in nutrient-poor or frozen environments, including the permafrost and oligotrophic wetlands [9]. The acetoclastic pathway involves the cleavage of acetate into CH_4 and CO_2, playing a central role in carbon-rich wetlands, rice paddies, and agricultural soils where acetate accumulates through plant residue decomposition [12]. In contrast, methylotrophic methanogenesis utilizes substrates like methanol or methylamines as electron donors and typically dominates under the saline or alkaline conditions [13]. These pathways differ biochemically but frequently coexist within the same environment, and their relative importance changes with factors such as temperature, pH, and nutrient levels. The varying prevalence of these pathways across ecosystems underscores how environmental conditions shape carbon flux and ultimately influence methane emissions. The last step in methanogenesis is catalyzed by the enzyme methyl-coenzyme M reductase (MCR), which is responsible for the formation of methane. The gene encoding this enzyme, *mcrA*, is widely regarded as the most reliable molecular marker for detecting both the presence and activity of methanogenic archaea [13, 14]. Molecular ecology studies show that *mcrA*-based quantitative PCR and metagenomic approaches reveal the strong correlations between methanogen abundance and methane emissions in different habitats [8]. As such, *mcrA* serves not only as a phylogenetic tool but also as a critical instrument for long-term monitoring of ecosystems sensitive to climate change.

Methanogenic archaea occupy the final step of anaerobic organic matter degradation through symbiotic interactions with fermentative bacteria and acetogens. This cooperation secures the thermodynamic sustainability of fermentation and maintains energy flow across ecosystems [9]. At the same time, methanogens face competition from sulfate-reducing bacteria, with the outcome of this balance shaped by the environmental variables such as substrate availability, pH, redox potential, and salinity [15]. For example, in rice paddies, pH and organic carbon levels are the decisive ecological factors that determine which methanogen groups will dominate [16, 17]. In permafrost thaw environments, hydrogenotrophic taxa often emerge first as pioneer colonizers, followed by a rise in acetoclastic groups [18, 19]. These observations show that methanogens are more than methane producers; their responses to climatic pressures play an active role in shaping ecosystem processes.

3. Diversity and distribution in soil ecosystems

At first glance, soil ecosystems may appear to be static matrices of mineral particles and organic matter. However, they are dynamic habitats where billions of microorganisms interact in complex ways. Among the most important engineers of this hidden world are methanogenic archaea. With their habitat-specific adaptation strategies and metabolic flexibility, they play a pivotal role in the global methane cycle. Recent advances in high-resolution metagenomic techniques and *mcrA*-based molecular monitoring have revealed that methanogen communities can rapidly restructure across both spatial and temporal scales [20–22]. For example, seasonal shifts in water levels in wetlands may lead to dramatic changes in community composition within just a few weeks.

The diversity of methanogenic archaea is shaped by a combination of physicochemical and climatic factors. Foremost among these is pH. In acidic environments, species composition changes markedly, often accompanied by shifts in the dominant methanogenic pathways [23]. Substrate availability is another decisive factor. The balance between acetate, hydrogen, and methyl compounds influences which metabolic pathway prevails. For instance, *Methanothrix* (formerly *Methanosaeta*) can thrive under low acetate concentrations, whereas *Methanosarcina* tends to dominate when acetate levels are high [24]. Water regime is equally critical, as it dictates the degree of anaerobiosis. A lowering of the groundwater table alters both redox conditions and substrate availability, triggering reorganization within the community [25]. Thermal tolerance also plays a defining role. Some hydrogenotrophic groups adapt to the elevated temperatures, while others function only within narrow thermal ranges [23]. These factors rarely act in isolation; rather, they interact. For example, during extended droughts, the persistence of hydrogenotrophic taxa helps sustain methane production, preventing a complete collapse of ecosystem functionality [26].

Methanogens express their diversity through distinct strategies across different habitats. In wetlands, for instance, the upper soil layers may contain oxygen at times, yet field data suggest that these layers can still account for up to 80% of net CH_4 flux [27]. In contrast, in arid regions, methanogenesis can become active in biological soil crusts even under light and oxygen exposure. In such cases, the choice of metabolic pathways shifts depending on whether conditions are oxic or anoxic [28]. These ecological examples make it clear that methanogens are not passive organisms confined to fixed niches. Instead, they are strategic actors capable of responding to environmental fluctuations and directly influencing the carbon cycle. Understanding how their diversity manifests across different ecological scenarios not only enriches basic microbial ecology but also provides essential insights for predicting climate response capacity and designing effective management strategies.

4. Responses to climate stressors and adaptation dynamics

As frozen soils in the Northern Hemisphere thaw or rice paddies in Asia undergo seasonal warming, methanogens respond rapidly to these environmental changes. Continuing climate change is reshaping methanogen communities, modifying their functional traits and their broader influence on ecosystem processes. Environmental factors such as temperature increase, elevated atmospheric CO_2, and cycles of drought and rewetting influence not only the amount of methane produced but also which methanogenic pathways become dominant. Each of these stressors imposes distinct selective pressures on methanogen communities, directly affecting ecosystem processes and greenhouse gas balances [10, 11, 23, 29]. **Table 1** summarizes the main methanogenic pathways, their representative Archaeal taxa, and how these groups respond to key environmental stressors such as temperature, moisture, and pH. It provides a concise overview that links biochemical mechanisms with ecological responses across diverse soil systems.

Of all stress factors, temperature and thermal tolerance are the strongest influences on methanogen activity. While mesophilic taxa typically thrive at 35–40°C, hyperthermophiles can remain metabolically active at temperatures above 80°C, and psychrophiles are capable of growth in the 15–25°C range. These physiological differences explain how temperature functions as a selective factor that influences both species distribution and the prevalence of specific methanogenic pathways within ecosystems. Experimental studies indicate that short-term warming tends to accelerate substrate turnover and CH_4 production, whereas long-term warming reduces diversity and favors a narrower set of dominant groups [23]. Field trials in Chinese rice paddies show that even a modest 2–3°C increase can favor hydrogenotrophic methanogens while reducing the abundance of acetoclastic groups [10]. However, it should be noted that while numerous studies report the associations between temperature, CO_2 enrichment, and the dominance of specific methanogenic pathways, these relationships are largely correlational rather than strictly causal, and further controlled experiments are needed to confirm the underlying mechanisms. Such findings suggest that under global warming scenarios, hydrogenotrophic pathways may increasingly take precedence.

Carbon dioxide enrichment and the resulting shifts in carbon availability act as equally powerful stressors. Elevated CO_2 levels enhance root exudation, driving more carbon into the soil. This often increases acetate availability, giving acetoclastic methanogens a competitive edge. Meta-analyses confirm that groups such as *Methanosarcinales* expand under these conditions [11]. When CO_2 enrichment occurs alongside warming, both methanogen abundance and methane production rates increase, further amplified by plant-microbe interactions in the rhizosphere [19, 30].

Repeated cycles of drought followed by rewetting create some of the most challenging environmental conditions for methanogens. During drought, anaerobic conditions collapse, and methanogen activity is strongly suppressed. Yet soil aggregates can maintain micro-anoxic niches, allowing hydrogenotrophic taxa to persist. This persistence provides the basis for rapid functional recovery once soils are rewetted. Field studies in the Zoige Plateau found that hydrogenotrophic methanogens

Methanogenic pathway	Representative Archaeal taxa	Primary substrates	Typical habitats	Response to key environmental stressors	Key references
Hydrogenotrophic	*Methanobacteriales*, *Methanococcales*, *Methanomicrobiales*	$H_2 + CO_2$	Wetlands, rice paddies, permafrost soils	Activity rises with moderate warming and H_2 availability but declines under drought due to moisture-driven redox shifts	[7, 8, 10, 24, 30]
Aceticlastic	*Methanosarcinales* (e.g., *Methanothrix* (=*Methanosaeta*), *Methanosarcina*)	Acetate	Peatlands, sediments, anaerobic digesters	Inhibited by temperature fluctuations and ammonium (NH_4^+) accumulation; more stable near neutral pH and steady moisture; gains competitive advantage when acetate accumulates.	[9, 11, 12, 24]
Methylotrophic	*Methanomassiliicoccales*, *Methanosarcinales*	Methanol, methylamines, methylsulfides	Marine/saline soils and sediments; disturbed or nutrient-poor environments	Often more tolerant to salinity and trace O_2; can become dominant under nutrient limitation or disturbance.	[13, 17, 31]
Hydrogen-dependent CO_2-reducing (mixotrophic)	*Methanocellales*, *Methanoregula*	$H_2 + CO_2$ ± organic acids	Paddy soils, thawing permafrost	Rapid recovery after drought→rewetting, contributing to post-stress CH_4 pulses; pathway prevalence shifts with temperature and moisture regimes.	[12, 29, 30]

Table 1.
Key methanogenic pathways, dominant archaea, and responses to environmental stressors.

maintained their dominance during drought and quickly regained activity upon rewetting, while acetoclastic taxa recovered more slowly [29]. These results indicate that hydrogenotrophic groups serve as a kind of biological insurance, sustaining ecosystem services under fluctuating climate pressures.

The *mcrA* gene, encoding the terminal enzyme of methanogenesis, remains the most reliable biomarker for evaluating methanogen responses to climate stressors. DNA copy numbers reflect community size, while RNA transcript abundance provides direct insight into metabolic activity. This dual signal makes it possible to detect functional changes even when overall community structure appears stable. Long-term thermal compensation effects described by Chen et al. (2020) or the rapid post-drought recovery observed by Li et al. (2022) were revealed only through time-resolved *mcrA* measurements [23, 29]. For this reason, *mcrA*-based monitoring has strategic value not only in basic ecological research but also in field restoration and applied biotechnology.

Methane has a global warming potential several times higher than that of CO_2 and an average atmospheric lifetime of about 12 years, giving it a lasting and significant impact on the planet's climate system. Incorporating methanogenic archaea into climate models is therefore essential for improving the accuracy of greenhouse gas projections. Modeling efforts must account not only for methanogen distribution but also for their dynamic responses to environmental variables such as temperature, moisture, pH, salinity, and substrate availability [28]. Data derived from *mcrA* provide robust calibration parameters, reflecting both active biomass and potential production capacity [32]. Recent syntheses further emphasize that hydrogenotrophic and acetoclastic pathways vary in dominance under the different environmental scenarios. Accounting for these differences will improve model accuracy and highlight the critical role of methanogens in the biosphere [24].

5. Biotechnological applications and climate solutions

Methanogenic archaea operate across a wide biotechnological spectrum from laboratory scale experiments to industrial bioreactors, and from degraded wetlands to restored ecosystems. As the final agents of organic matter conversion into methane under anaerobic conditions, they hold strategic importance for both renewable energy production and greenhouse gas management. In the context of climate change, their applications extend beyond enhancing renewable energy potential; they also offer novel pathways for mitigating methane emissions and stabilizing the carbon cycle [8, 33].

Industrial anaerobic digesters are the most widely used platforms for converting organic waste into methane. In these systems, acetoclastic taxa such as *Methanothrix* (formerly *Methanosaeta*, family *Methanosaetaceae*) maintain steady methane output under high acetate concentrations, while hydrogenotrophic groups such as *Methanobacteriales* and *Methanomicrobiales* convert H_2 and CO_2 into methane, ensuring overall system stability [33]. In recent years, microaeration strategies have been developed to prevent oxidative stress-tolerant species from becoming overly dominant, thereby improving the biogas yields. Additionally, immobilizing methanogens on carrier surfaces has been shown to increase cell density and reduce reactor startup time [34].

Bioaugmentation involves the deliberate introduction of stress-tolerant methanogenic strains into existing systems to enhance the process efficiency. For example, continuous stirred tank reactor experiments under ammonia-rich conditions demonstrated that adding fast-growing hydrogenotrophic *Methanoculleus bourgensis* MS2(T) increased methane yields by more than 30% [35]. Similarly, batch anaerobic digestion trials with chicken manure showed that applying selected methanogenic cultures improved methane production and shortened the digestion period [36]. Targeted enrichment with hydrogenotrophic methanogens has also resulted in the modest but consistent gains, such as a cumulative 4% increase in CH_4 yields [37]. When combined with the biochar amendments, these strategies accelerated methane production and improved reactor stability, with reported increases in methane yields compared to non-augmented controls [38–40]. In addition to strain introduction, immobilizing methanogens on carrier materials has proven effective in increasing cell density and reducing reactor start-up time. Biofilm-based systems and biochar carriers have been shown to facilitate the faster community establishment and improve methane production rates [38–40]. Biorestoration strategies, on the other hand, aim to reintroduce native methanogenic communities into degraded wetlands and peatlands. Long-term rewetting experiments in European peatlands have shown significant increases in *mcrA* gene abundance and functional recovery of methanogenic activity, demonstrating that restoration practices can reinstate ecological functionality and stabilize greenhouse gas fluxes [41, 42]. The *mcrA* gene is a key biomarker for monitoring methanogenic processes. At the reactor scale, *mcrA* copy number can serve as an early-warning indicator of system stability, while time-series analyses of *mcrA* transcripts at the field scale directly reflect responses to climate stressors [14]. Because DNA reflects community size and RNA indicates metabolic activity, evaluating both layers provides a reliable method for assessing the success of biotechnological applications.

Methanogen-based technologies offer a dual climate benefit: producing renewable biomethane to reduce reliance on fossil fuels, and capturing and utilizing methane to lower net emissions. Looking forward, genetic editing tools such as CRISPR-Cas are expected to improve the stress tolerance of methanogens and broaden their substrate utilization spectrum [43]. Such advances could enhance climate-friendly biotechnology strategies and facilitate the integration of methanogens into the sustainable development goals. During anaerobic decomposition, fermentative and acetogenic stages generate acetate, formate, H_2, and CO, which serve as substrates for methanogens. Hydrogenotrophic taxa consume H_2, maintaining the thermodynamic balance of fermentation, while acetoclastic taxa utilize acetate [8]. When hydrogen partial pressure rises, hydrogenotrophic methanogens gain an advantage, whereas acetoclastic groups dominate under low H_2 conditions [7]. Sulfate-reducing bacteria can outcompete methanogens, suppressing methanogenesis in some environments [44]. Methylotrophic methanogens further contribute by metabolizing compounds such as dimethyl sulfide and trimethylamine, thereby linking the sulfur and carbon cycles [45]. Shifts in community composition often occur during new habitat colonization or under climate stressors. For example, following permafrost thaw, wetlands are initially dominated by hydrogenotrophic species, but over time acetoclastic taxa increase in prevalence [31, 46]. These successional dynamics are tightly linked to substrate availability and hydrological regimes, which in turn strongly influence CH_4 emissions [23].

In natural systems, the overarching goal is to reduce excessive methane emissions. Strategies include regulating hydrological regimes, balancing substrate supply, and minimizing stressors such as extreme pH, salinity, or toxic metals [47, 48]. Although rewetting can temporarily increase emissions, long-term equilibrium may still be achieved [49, 50]. Implementing such interventions in conjunction with *mcrA*-based monitoring provides the critical tools for greenhouse gas management [51, 52]. In drying soils, consortia of microalgae and fungi can create millimeter-scale oxic-anoxic mosaics, restructuring methanogenic niches. Field-based evidence supports this integration of microbial monitoring and restoration policy. In three long-term peatland rewetting sites in northern Germany, the study demonstrated that mcrA-based qPCR tracking effectively captured the recovery of methanogenic communities, providing a scientific basis for water-level management decisions [41]. Similarly, another study showed that peatland restoration following mining-altered methane-cycling potential and methanogen composition, emphasizing the need to incorporate microbial indicators into greenhouse gas mitigation frameworks [42]. Together, these studies highlight how applied microbial ecology can guide real-world management and strengthen the connection between research and policy. Algal exudates and fungal hyphae form microenvironments that alter substrate availability and redox conditions, thereby supporting either hydrogenotrophic or acetoclastic pathways [28, 53]. When tracked through *mcrA* time-series analyses, such interactions may represent complementary biotechnological approaches for managing greenhouse gas emissions.

6. Future perspectives and sustainability

The ecological and biotechnological dynamics discussed in the previous sections offer critical insights for sustainability policies. The global methane cycle is shaped not only by methanogenic archaea, the sole biological producers of methane, but also by methanotrophic bacteria, which oxidizes CH_4 and partially buffer emissions. In wetlands, rice paddies, and soil ecosystems, methanotrophs act as counterbalancing agents, helping to maintain methane equilibrium [54, 55]. Still, the focus here remains on methanogenic archaea, given their direct role in methane production processes and their potential leverage within climate and biotechnology frameworks.

Methanogens occupy a dual position in the climate change narrative. As the only biological producers of methane under anaerobic conditions, they contribute directly to the rise of atmospheric greenhouse gases. Methane, with its far greater short-term warming potential compared to carbon dioxide, can accelerate the course of the climate crisis depending on methanogenic activity [8]. At the same time, these microorganisms are part of the solution: When harnessed in controlled biotechnological systems, methane can be used as a renewable energy source, and ecosystem-based interventions, such as wetland and peatland restoration, can help regulate methane fluxes [46, 56]. This dual influence positions methanogens not only as contributors to the problem but also as strategic actors in sustainability policies.

Climate stressors such as rising temperatures, shifting moisture regimes, and permafrost thaw exert direct impacts on the structure of methanogenic communities and the distribution of their metabolic pathways [23]. Managing these interactions effectively in the coming decades will be central to sustainable energy production, ecosystem restoration, and climate adaptation policies. In boreal regions, hydrogenotrophic taxa such as *Methanoregula* and *Methanocella* gain strategic importance due to their cold tolerance, while in temperate zones, acetoclastic groups like *Methanothrix* and *Methanosarcina* play more dominant roles [57]. Advances in multilayered bio-observation techniques, including metagenomics, metatranscriptomics, and isotope-based methods have made it possible to identify the climate-sensitive microbial taxa.

These tools provide a critical foundation for selecting strains in biotechnology and informing adaptive management in field restoration projects [46, 58].

Moreover, to ensure broader geographic representation, field studies from North America have also provided critical insights into methanogen-environment interactions. For instance, a comprehensive survey of 17 peatland sites across eastern North America demonstrated that trace metal concentrations (particularly Ni and Cu) strongly predicted the methanogenic community composition through mcrA gene profiling [59]. These findings emphasize that metagenomic monitoring can support restoration and greenhouse gas management strategies beyond European and Asian contexts.

Emerging research frontiers further demonstrate how these approaches reshape our understanding of methanogen ecology and application. *Metatranscriptomic* analyses in Arctic peat soils have revealed that active methanogenic pathways shift dynamically with temperature and redox changes, allowing real-time tracking of functional responses to climate stressors [46]. Complementary isotopic tracing studies in permafrost environments have separated the contributions of hydrogenotrophic and acetoclastic pathways to methane flux, demonstrating the mechanistic linkage between microbial metabolism and greenhouse gas emissions [31]. On the biotechnological side, *CRISPR-Cas9–enabled genome editing* in model methanogens such as *Methanosarcina acetivorans* has opened new possibilities for improving stress tolerance, substrate utilization, and methane yield efficiency [43]. Together, these frontier methods bridge the fundamental microbiology with climate-adaptive biotechnology, pointing toward precision approaches for methane cycle management.

Methanogen-based strategies also align directly with the United Nations' Sustainable Development Goals (SDGs). Renewable biomethane production reduces dependence on fossil fuels, reinforcing energy security (SDG 7). Controlled methane management contributes to lowering greenhouse gas emissions (SDG 13). Wetland and peatland restoration supports the functioning of terrestrial ecosystems and safeguards soil biodiversity (SDG 15). Thus, methanogens should not be regarded merely as a subject of microbial ecology, but also as key components of technological solutions that intersect with climate policy.

Recent global initiatives further highlight the policy relevance of methane mitigation. The UNEP Eye on Methane 2024 report emphasizes that data-driven monitoring and rapid emission reduction could cut global warming by up to 0.1°C within two decades, noting that "the tools for a methane data revolution are already in place [60]." Similarly, the IEA Global Methane Tracker 2025 presents country-level emission maps and identifies that roughly 70% of methane emissions from fossil fuel operations can be avoided using the existing technologies at low or no net cost [60, 61]. Integrating such frameworks with microbial monitoring, particularly *mcrA*-based ecological assessments, can bridge global policy efforts with on-the-ground ecosystem management.

Nonetheless, significant questions remain unresolved despite the rapid growth of research. The metabolic adaptations of species thriving in extreme environments such as acidic peatlands, high-salinity soils, or heavy metal-polluted sites remain poorly understood. The ecological consequences of methanogens' symbiotic interactions with other archaea and anaerobic bacteria are also not fully clear. Moreover, there is still limited knowledge about the long-term ecological effects of genetic engineering and CRISPR-Cas-based modifications [8, 43, 62, 63]. Addressing these gaps could yield breakthroughs in both carbon cycle management and applied biotechnology.

Methanogenic archaea are likely to become integral not only at the laboratory scale but also in global climate governance, ecosystem restoration strategies, and sustainability policies. A holistic understanding of their ecological roles, biotechnological potential, and interplay with climate frameworks will be decisive for the success of future climate solutions.

7. Conclusions

Methanogenic archaea are strategic microorganisms whose unique roles in the global carbon cycle position them both as potential accelerators of climate change and as central players in sustainable energy production. As the sole biological source of methane in anaerobic ecosystems, they carry out this function through hydrogenotrophic, acetoclastic, and methylotrophic pathways. These biochemical processes can be precisely tracked *via* the *mcrA* gene, enabling direct linkage between community shifts, functional capacity, and environmental stressors such as climate variability [14, 64].

Their diversity in soil ecosystems acts as a form of biological insurance, safeguarding ecosystem services against disruption. Adaptive responses to stressors, including rising temperatures, CO_2 enrichment, drought, and rewetting cycles, make methanogens critical actors in climate adaptation strategies [23, 29]. Patterns of species distribution across wetlands, permafrost regions, and agricultural soils provide direct ecological data that can inform future climate-responsive management strategies.

At the biotechnological level, methanogens offer a dual advantage. On the one hand, renewable biomethane production can reduce dependence on fossil fuels; on the other, controlled methane capture and utilization can lower net greenhouse gas emissions [56, 57]. This duality highlights that methanogens are not merely part of the climate problem but also hold the keys to its solution. Applications such as wetland restoration, organic waste management, biogas systems, and emerging "power-to-gas" technologies integrate their ecological functions with technological potential in a complementary manner.

Recent studies underscore that *mcrA*-based bio-observation provides a powerful way to track methanogen responses to climate stressors, with applications ranging from improving biogas efficiency to guiding biological restoration of degraded ecosystems [1, 3, 8]. Yet, significant gaps remain. The metabolic adaptations of taxa inhabiting extreme environments, such as acidic peatlands, hypersaline soils, or heavy metal contaminated sites, are not yet fully understood. Likewise, long-term ecological effects of genetic engineering and CRISPR-Cas-based modifications remain largely unexplored [62, 63]. Closing these gaps will be vital for both effective carbon cycle management and the safe integration of biotechnological applications.

In conclusion, methanogenic archaea should be seen not only as the subjects of microbial ecology but also as indispensable components of global climate governance, ecosystem restoration, and sustainability policy. Aligned with the UN Sustainable Development Goals (SDG 7, 13, 15), they are expected to play a key role in the transition to clean energy and in protecting terrestrial ecosystems. Research conducted at the intersection of ecology, molecular biology, and engineering will deepen our understanding of their diversity, functionality, and adaptive capacity, while directly contributing to innovative and sustainable climate solutions. Thus, methanogenic archaea are repositioned not just as the invisible actors of the climate crisis, but as strategic partners in sustainable energy, ecosystem restoration, and climate policy. Strengthening our knowledge of their ecological functions and integrating them into biotechnological systems will be among the decisive steps in shaping the planet's future.

Acknowledgements

The author acknowledges the use of ChatGPT for language editing in the chapter.

Notes/thanks/other declarations

This work was supported by the Scientific Research Projects Coordination Unit of Istanbul University (Project No. FDK-2024-40669).

Author details

Gamze Gürerk and Sevcan Aydin*
Division of Biotechnology, Biology Department, Faculty of Science, Istanbul University, Istanbul, Turkey

*Address all correspondence to: sevcan.aydin@istanbul.edu.tr

References

[1] (IPCC) IP on CC. Climate Change 2021 – The Physical Science Basis: Working Group I Contribution to the Sixth Assessment Report of the Intergovernmental Panel on Climate Change. Climate Change 2021 – The Physical Science Basis. Available from: https://www.cambridge.org/core/books/climate-change-2021-the-physical-science-basis/415F29233B8BD19FB55F65E3DC67272B

[2] Etminan M, Myhre G, Highwood EJ, Shine KP. Radiative forcing of carbon dioxide, methane, and nitrous oxide: A significant revision of the methane radiative forcing. Geophysical Research Letters. 2016;**43**(24):12614-12623. DOI: 10.1002/2016GL071930

[3] Trends in CH4 - NOAA Global Monitoring Laboratory. Available from: https://gml.noaa.gov/ccgg/trends_ch4/

[4] Dlugokencky EJ, Steele LP, Lang PM, Masarie KA. The growth rate and distribution of atmospheric methane. Journal of Geophysical Research. 1994;**99**(D8):17021-17043. DOI: 10.1029/94JD01245

[5] Dlugokencky EJ, Houweling S, Bruhwiler L, Masarie KA, Lang PM, Miller JB, et al. Atmospheric methane levels off: Temporary pause or a new steady-state? Geophysical Research Letters. 2003;**30**(19):1826. DOI: 10.1029/2003GL018126

[6] Zhao Y, Saunois M, Bousquet P, Lin X, Hegglin MI, Canadell JG, et al. Reconciling the bottom-up and top-down estimates of the methane chemical sink using multiple observations. Atmospheric Chemistry and Physics. 2023;**23**:789-807. DOI: 10.5194/acp-23-789-2023

[7] Thauer RK, Kaster AK, Seedorf H, Buckel W, Hedderich R. Methanogenic archaea: Ecologically relevant differences in energy conservation. Nature Reviews Microbiology. 2008;**6**:579-591

[8] Conrad R. Importance of hydrogenotrophic, aceticlastic and methylotrophic methanogenesis for methane production in terrestrial, aquatic and other anoxic environments: A mini review. Pedosphere. 2020;**30**(1):25-39. Available from: https://www.sciencedirect.com/science/article/abs/pii/S1002016018600529

[9] Evans PN, Boyd JA, Leu AO, Woodcroft BJ, Parks DH, Hugenholtz P, et al. An evolving view of methane metabolism in the archaea. Nature Reviews Microbiology. 2019;**17**(4):219-232. Available from: https://pubmed.ncbi.nlm.nih.gov/30664670/

[10] Mo D, Liu Y, Li M, Li H, Zhang J, Qian H, et al. Warming increases CH_4 emissions from rice paddies through shifts in methanogenic and methanotrophic communities. Soil Biology and Biochemistry. 2025;**200**:109639. Available from: https://www.sciencedirect.com/science/article/abs/pii/S0038071724003286

[11] Ding Y, Wang M, Du X, He X, Xu T, Liu X, et al. The impact of elevated CO_2 on methanogen abundance and methane emissions in terrestrial ecosystems: A meta-analysis in brief the impact of elevated CO_2 on methanogen abundance and methane emissions in terrestrial ecosystems: A meta-analysis. iScience. 2024;**27**:111504. DOI: 10.1016/j.isci.2024.111504

[12] Lyu Z, Shao N, Akinyemi T, Whitman WB. Methanogenesis. Current Biology. 2018;**28**(13):R727-R732.

Available from: https://pubmed.ncbi.nlm.nih.gov/29990451/

[13] Luton PE, Wayne JM, Sharp RJ, Riley PW. The mcrA gene as an alternative to 16S rRNA in the phylogenetic analysis of methanogen populations in landfill. Microbiology. 2002;**148**(11):3521-3530. DOI: 10.1099/00221287-148-11-3521

[14] Steinberg LM, Regan JM. mcrA-targeted real-time quantitative PCR method to examine methanogen communities. Applied and Environmental Microbiology. 2009;**75**:4435-4442

[15] Zu Q, Zhong L, Deng Y, Shi Y, Wang B, Jia Z, et al. Geographical distribution of methanogenic archaea in nine representative paddy soils in China. Frontiers in Microbiology. 2016;7(SEP):187506. Available from: www.frontiersin.org

[16] Angel R, Claus P, Conrad R. Methanogenic archaea are globally ubiquitous in aerated soils and become active under wet anoxic conditions. The ISME Journal. 2012;**6**(4):847-862. DOI: 10.1038/ismej.2011.141

[17] Asakawa S. Ecology of methanogenic and methane-oxidizing microorganisms in paddy soil ecosystem. Soil Science & Plant Nutrition. 2021;**67**(5):520-526. DOI: 10.1080/00380768.2021.1953355

[18] Bischoff J, Mangelsdorf K, Gattinger A, Schloter M, Kurchatova AN, Herzschuh U, et al. Response of methanogenic archaea to Late Pleistocene and Holocene climate changes in the Siberian Arctic. Global Biogeochemical Cycles. 2013;**27**(2):305-317. DOI: 10.1029/2011GB004238

[19] Weiland-Bräuer N. Friends or foes—Microbial interactions in nature. Biology (Basel). 2021;**10**(6):496. DOI: 10.3390/biology10060496

[20] Yan W, Wang D, Wang Y, Wang C, Chen X, Liu L, et al. Metatranscriptomics-guided genome-scale metabolic reconstruction reveals the carbon flux and trophic interaction in methanogenic communities. Microbiome. 2024;**12**(1):1-18. DOI: 10.1186/s40168-024-01830-z

[21] Nawarathna TNTK, Fujii N, Yamamoto K, Kuroda K, Narihiro T, Ozaki N, et al. Metagenomic insights into Candidatus Scalindua in a long-term cultivated marine Anammox consortium: The important role of tetrahydrofolate-mediated carbon fixation. Microbes and Environments. 2025;**40**(2):ME25007. Available from: https://pmc.ncbi.nlm.nih.gov/articles/PMC12213060/

[22] Zhang Y, Jing H. Metagenome sequencing and 107 microbial genomes from seamount sediments along the Yap and Mariana trenches. Scientific Data. 2024;**11**:887. DOI: 10.1038/s41597-024-03762-7

[23] Chen H, Zhu T, Li B, Fang C, Nie M. The thermal response of soil microbial methanogenesis decreases in magnitude with changing temperature. Nature Communications. 2020;**11**(1):1-7. Available from: https://www.nature.com/articles/s41467-020-19549-4

[24] Liu Y, Whitman WB. Metabolic, phylogenetic, and ecological diversity of the methanogenic archaea. Annals of the New York Academy of Sciences. 2008;**1125**(1):171-189. DOI: 10.1196/annals.1419.019

[25] Watanabe T, Kimura M, Asakawa S. Community structure of methanogenic archaea in paddy field soil under double cropping (rice–wheat). Soil Biology

and Biochemistry. 2006;**38**(6):1264-1274. Available from: https://www.sciencedirect.com/science/article/abs/pii/S0038071705003408

[26] He C, Han T, Tan L, Li X. Effects of dark septate endophytes on the performance and soil microbia of lycium ruthenicum under drought stress. Frontiers in Plant Science. 2022;**13**:898378. DOI: 10.3389/fpls.2022.898378

[27] Wilmoth JL, Schaefer JK, Schlesinger DR, Roth SW, Hatcher PG, Shoemaker JK, et al. The role of oxygen in stimulating methane production in wetlands. Global Change Biology. 2021;**27**(22):5831. Available from: https://pmc.ncbi.nlm.nih.gov/articles/PMC9291790/

[28] Angel R, Matthies D, Conrad R. Activation of methanogenesis in arid biological soil crusts despite the presence of oxygen. PLoS One. 2011;**6**(5):e20453. DOI: 10.1371/journal.pone.0020453

[29] Li W, Shi R, Yuan L, Lan X, Feng D, Chen H. Effect of short-term warming and drought on the methanogenic communities in degraded peatlands in Zoige Plateau. Frontiers in Microbiology. 2022;**13**:880300

[30] Shakoor A, Pendall E, Macdonald CA. Microbial mechanisms of interactive climate-driven changes in soil N_2O and CH_4 fluxes: A global meta-analysis. Journal of Environmental Management. 2025;**376**:124380. Available from: https://www.sciencedirect.com/science/article/pii/S0301479725003561

[31] McCalley CK, Woodcroft BJ, Hodgkins SB, Wehr RA, Kim EH, Mondav R, et al. Methane dynamics regulated by microbial community response to permafrost thaw. Nature. 2014;**514**(7523):478-481. Available from: https://www.nature.com/articles/nature13798

[32] Cui S, Guo H, Pugliese L, Nielsen CK, Wu S. Controlled burning of peat before rewetting modifies soil chemistry and microbial dynamics to reduce short-term methane emissions. Communications Earth & Environment. 2025;**6**(1):1-12. Available from: https://www.nature.com/articles/s43247-025-02336-8

[33] Hori T, Haruta S, Ueno Y, Ishii M, Igarashi Y. Dynamic transition of a methanogenic population in response to the concentration of volatile fatty acids in a thermophilic anaerobic digester. Applied and Environmental Microbiology. 2006;**72**(2):1623-1630. Available from: https://pubmed.ncbi.nlm.nih.gov/16461718/

[34] González R, Cabeza IO, Casallas-Ojeda M, Gómez X. Biological hydrogen methanation with carbon dioxide utilization: Methanation acting as mediator in the hydrogen economy. Environments. 2023;**10**(5):82. Available from: https://www.mdpi.com/2076-3298/10/5/82/htm

[35] Fotidis IA, Wang H, Fiedel NR, Luo G, Karakashev DB, Angelidaki I. Bioaugmentation as a solution to increase methane production from an ammonia-rich substrate. Environmental Science & Technology. 2014;**48**(13):7669-7676. DOI: 10.1021/es5017075

[36] Linsong H, Lianhua L, Ying L, Changrui W, Yongming S. Bioaugmentation with methanogenic culture to improve methane production from chicken manure in batch anaerobic digestion. Chemosphere. 2022;**303**(Pt 3). Available from: https://pubmed.ncbi.nlm.nih.gov/35654231/

[37] Gállego-Bravo AK, García-Mena J, Piña-Escobedo A, López-Jiménez G,

Gutiérrez-Castillo ME, Tovar-Gálvez LR. Monitoring of a microbial community during bioaugmentation with hydrogenotrophic methanogens to improve methane yield of an anaerobic digestion process. Biotechnology Letters. 2023;**45**(10):1339-1353. Available from: https://pubmed.ncbi.nlm.nih.gov/37535136/

[38] Yang HJ, Yang ZM, Xu XH, Guo RB. Increasing the methane production rate of hydrogenotrophic methanogens using biochar as a biocarrier. Bioresource Technology. 2020;**302**:122829. Available from: https://www.sciencedirect.com/science/article/abs/pii/S0960852420300985

[39] Rachbauer L, Voitl G, Bochmann G, Fuchs W. Biological biogas upgrading capacity of a hydrogenotrophic community in a trickle-bed reactor. Applied Energy. 2016;**180**:483-490. Available from: https://www.sciencedirect.com/science/article/abs/pii/S0306261916310388

[40] Habouzit F, Hamelin J, Santa-Catalina G, Steyer JP, Bernet N. Biofilm development during the start-up period of anaerobic biofilm reactors: The biofilm archaea community is highly dependent on the support material. Microbial Biotechnology. 2014;**7**(3):257-264. DOI: 10.1111/1751-7915.12115

[41] Weil M, Wang H, Bengtsson M, Köhn D, Günther A, Jurasinski G, et al. Long-term rewetting of three formerly drained peatlands drives congruent compositional changes in pro- and eukaryotic soil microbiomes through environmental filtering. Microorganisms. 2020;**8**(4):550. DOI: 10.3390/microorganisms8040550

[42] Reumer M, Harnisz M, Lee HJ, Reim A, Grunert O, Putkinen A, et al. Impact of peat mining and restoration on methane turnover potential and methane-cycling microorganisms in a northern bog. Applied and Environmental Microbiology. 2018;**84**(3):e02218-17. DOI: 10.1128/AEM.02218-17

[43] Li J, Zhang L, Xu Q, Zhang W, Li Z, Chen L, et al. CRISPR-Cas9 toolkit for genome editing in an autotrophic CO_2-fixing methanogenic archaeon. Microbiology Spectrum. 2022;**10**(4). DOI: 10.1128/spectrum.01165-22?download=true

[44] Muyzer G, Stams AJM. The ecology and biotechnology of sulphate-reducing bacteria. Nature Reviews Microbiology. 2008;**6**(6):441-454. Available from: https://www.nature.com/articles/nrmicro1892

[45] Borrel G, O'Toole PW, Harris HMB, Peyret P, Brugère JF, Gribaldo S. Phylogenomic data support a seventh order of methylotrophic methanogens and provide insights into the evolution of methanogenesis. Genome Biology and Evolution. 2013;**5**(10):1769-1780. DOI: 10.1093/gbe/evt128

[46] Tveit AT, Urich T, Svenning MM. Metatranscriptomic analysis of arctic peat soil microbiota. Applied and Environmental Microbiology. 2014;**80**(18):5761-5772. DOI: 10.1128/aem.01030-14?download=true

[47] Juottonen H. Disentangling the effects of methanogen community and environment on peatland greenhouse gas production by a reciprocal transplant experiment. Functional Ecology. 2020;**34**(6):1268-1279. DOI: 10.1111/1365-2435.13536

[48] Putkinen A, Larmola T, Tuomivirta T, Siljanen HMP, Bodrossy L, Tuittila ES, et al. Peatland succession

induces a shift in the community composition of *Sphagnum*-associated active methanotrophs. FEMS Microbiology Ecology. 2014;**88**(3):596-611. DOI: 10.1111/1574-6941.12327

[49] Orejuela-Escobar L, Gualle A, Ochoa-Herrera V, Philippidis GP. Prospects of microalgae for biomaterial production and environmental applications at biorefineries. Sustainability (Switzerland). 2021;**13**(6):1-19

[50] Abdalla M, Hastings A, Truu J, Espenberg M, Mander Ü, Smith P. Emissions of methane from northern peatlands: A review of management impacts and implications for future management options. Ecology and Evolution. 2016;**6**(19):7080-7102. DOI: 10.1002/ece3.2469

[51] Dusengemungu L, Gwanama C, Simuchimba G, Mubemba B. Potential of bioaugmentation of heavy metal contaminated soils in the Zambian Copperbelt using autochthonous filamentous fungi. Frontiers in Microbiology. 2022;**13**:1045671

[52] Lacroix EM, Aeppli M, Boye K, Brodie E, Fendorf S, Keiluweit M, et al. Consider the anoxic microsite: Acknowledging and appreciating spatiotemporal redox heterogeneity in soils and sediments. ACS Earth and Space Chemistry. 2023;**7**(9):1592. Available from: https://pmc.ncbi.nlm.nih.gov/articles/PMC10519444/

[53] Angle JC, Morin TH, Solden LM, Narrowe AB, Smith GJ, Borton MA, et al. Methanogenesis in oxygenated soils is a substantial fraction of wetland methane emissions. Nature Communications. 2017;**8**(1):1-9. Available from: https://www.nature.com/articles/s41467-017-01753-4

[54] Hanson RS, Hanson TE. Methanotrophic bacteria. Microbiological Reviews. 1996;**60**(2):439-471. DOI: 10.1128/mr.60.2.439-471.1996

[55] Knief C. Diversity and habitat preferences of cultivated and uncultivated aerobic methanotrophic bacteria evaluated based on pmoA as molecular marker. Frontiers in Microbiology. 2015;**6**(DEC):170928. Available from: www.frontiersin.org

[56] Juottonen H, Hynninen A, Nieminen M, Tuomivirta TT, Tuittila ES, Nousiainen H, et al. Methane-cycling microbial communities and methane emission in natural and restored peatlands. Applied and Environmental Microbiology. 2012;**78**(17):6386-6389. DOI: 10.1128/aem.00261-12?download=true

[57] Tveit AT, Söllinger A, Rainer EM, Didriksen A, Hestnes AG, Motleleng L, et al. Thermal acclimation of methanotrophs from the genus *Methylobacter*. ISME Journal. 2023;**17**:502-513

[58] Mateos DM, Bhatnagar JM. Restoring ecological complexity in a changing environment. Current Biology. 2024;**34**(9):R365-R371. Available from: https://www.cell.com/action/showFullText?pii=S0960982224003993

[59] Bear SE, Seward JD, Lamit LJ, Basiliko N, Moore T, Lilleskov E, et al. Beyond the usual suspects: Methanogenic communities in eastern North American peatlands are also influenced by nickel and copper concentrations. FEMS Microbiology Letters. 2021;**368**(21):24. DOI: 10.1093/femsle/fnab151

[60] UNEP - UN Environment Programme. Available from: https://www.unep.org/interactives/eye-on-methane-2024/

[61] Da Diamond D. Global Methane Tracker. 2025. Available from: www.iea.org

[62] Myers T, Dykstra CM. Teaching old dogs new tricks: Genetic engineering methanogens. Applied and Environmental Microbiology. 2024;**90**(7):e02247-23. DOI: 10.1128/aem.02247-23

[63] Zhu P, Somvanshi T, Bao J, Scheller S. CRISPR/Cas12a toolbox for genome editing in *Methanosarcina acetivorans*. Frontiers in Microbiology. 2023;**14**:1235616

[64] Cui M, Ma A, Qi H, Zhuang X, Zhuang G, Zhao G. Warmer temperature accelerates methane emissions from the Zoige wetland on the Tibetan Plateau without changing methanogenic community composition. Scientific Reports. 2015;**5**:11616. DOI: 10.1038/srep11616